FROG DISSECTION MANUAL

Bruce D. Wingerd

Illustrated by Geoffrey Stein, D.V.M.

THE JOHNS HOPKINS UNIVERSITY PRESS
Baltimore & London

© 1988 The Johns Hopkins University Press
All rights reserved
Printed in the United States of America

The Johns Hopkins University Press
701 West 40th Street
Baltimore, Maryland 21211
The Johns Hopkins Press Ltd., London

The paper used in this publication meets the minimum require-
ments of American National Standard for Information Sciences
—Permanence of Paper for Printed Library Materials, ANSI
Z39.48-1984.
ISBN 0-8018-3601-8 (pbk.: alk. paper)

Contents

Contents

Illustrations

Introduction

FOR THE PAST two hundred years or more the frog has been given the dubious honor of having been the favorite subject for biological experimentation. Its popularity as a specimen has historically been the result of a wide distribution, ease of capture and maintenance, and ability to survive experimental situations that would preclude the use of larger animals. Early investigators used frogs extensively in their efforts to unravel the mysteries of the heart, blood circulation, nerve stimulation, and muscle contraction. In more recent times, the use of frogs in science has narrowed to classroom demonstrations of heart and muscle function in living or freshly sacrificed specimens and general dissection of preserved specimens for the study of body structure.

The use of the frog as a dissection specimen affords beginning students a firsthand look at a vertebrate that has many body structures in a position and arrangement similar to that of the human. The experience has merit, for dissection is a unique, "hands-on" approach to the study of biology. Going beyond the visual investigation of textbook drawings and photographs, dissection allows the student to examine real structures, using the additional sense of touch. This practical approach has the potential for opening up new dimensions of learning if it is taught properly. It is the purpose of this manual to present dissection in a clear, step-by-step manner in order to make the learning experience meaningful and effective.

The frog is a member of the class Amphibia, which represents a group between animals that are completely aquatic and those that are completely terrestrial. Although they have legs for land locomotion, they must keep close to water or moist soil to prevent critical water loss. Amphibians, as a rule, are ectothermic, that is, their body temperature and rate of metabolism are dependent on ambient temperature.

There is only one genus of true frog: *Rana*. The most common frog in the United States is the leopard frog, *Rana pipiens*. This abundant species is a popular subject in biology experiments and in dissection, but another species, the bullfrog, *Rana catesbeiana*, may be preferable in dissection owing to its larger size. All drawings in

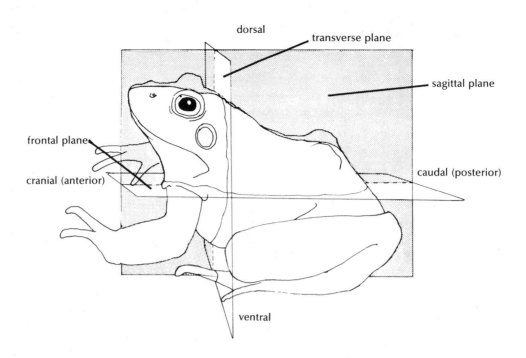

FIGURE N.1. Descriptive terminology and planes of section

this manual are taken from bullfrog subjects. The taxonomic classification of the bullfrog is as follows:

Phylum: Chordata
Subphylum: Vertebrata
Class: Amphibia
Order: Anura
Family: Batrachia
Genus: *Rana*
Species: *catesbeiana*

Before beginning your study of the frog, study the following directional and spatial terms. Note that these terms apply primarily to quadrupeds, or four-legged animals. This introductory step is essential because these terms will be used extensively throughout the text (Fig. N.1).

Directional Terms

Rostral: toward the nose end.
Cranial/anterior: toward the head end.
Caudal/posterior: toward the tail end.
Dorsal: toward the back side.
Ventral: toward the belly side.
Midline: an imaginary plane that bisects the body into right and left halves.

Median: lying in or near the midline.
Medial: lying closer to the midline relative to another structure.
Lateral: lying farther away from the midline relative to another structure.
Proximal: near a structure's origin or point of attachment to the body.
Distal: away from a structure's origin or point of attachment to the body.
Superficial: toward the body surface.
Deep: away from the body surface.

Planes of Section

Transverse (cross): a plane that passes at a right angle to the long axis of a body or body structure, usually resulting in cranial and caudal portions.
Longitudinal: a plane that extends from cranial to caudal along the long axis of the body; the longitudinal plane bisects the transverse plane at a right angle.
Sagittal: a longitudinal plane that divides the body into right and left halves; if this division is into equal halves, it is called **midsagittal**. If it is into unequal halves, it is called **parasagittal**.
Frontal (coronal): a longitudinal plane that extends from cranial to caudal and horizontally from right to left, dividing the body into ventral and dorsal portions.

External Anatomy & the Skin

EXTERNAL ANATOMY

PLACE YOUR SPECIMEN on a dissecting pan and examine its external features (Figs. 1.1, 1.2). The body of the frog is divided into two portions, the **head** and the **trunk**. Unlike mammals, the head is attached directly to the trunk without a neck, and there is no tail. Now try to move the head while keeping the trunk stationary with your fingers. The inability of the frog to pivot its head without moving its entire body as you have just tried to do places an important limitation on the frog's ability to sense efficiently its external environment. However, it seems to increase its stability when jumping.

HEAD

Examine the head region of your frog more closely. Notice that the frog contains two large spherical **eyes**, each with a fixed **upper lid** and a thinner **lower lid**. Also associated with each eye is a transparent **nictitating membrane**, which can move upward over the eyeball for its protection. Located at the end of the snout are two small openings through which air flows during breathing, called the **nostrils** or **external nares**. The wide **mouth** can be seen on the ventral side of the snout. Dorsal to the corner of the mouth on each side of the head is a round disc called the **tympanum**. This is a membrane that receives sound waves for hearing.

TRUNK

Attached to the trunk are four appendages, the two **forelimbs** and two **hind limbs**. Each forelimb contains four **digits**, and each hind limb contains five **webbed digits**. Between the hind limbs is the **cloacal aperture**, which opens into the **cloaca**. The cloaca is a chamber exclusive to most fish, amphibians, reptiles, birds, and monotreme mammals that serves as a collection vessel for solid wastes, liquid wastes, and sex cells from the reproductive organs.

THE SKIN

The skin of the frog is an extremely dynamic organ. It functions as protection for underlying structures and from

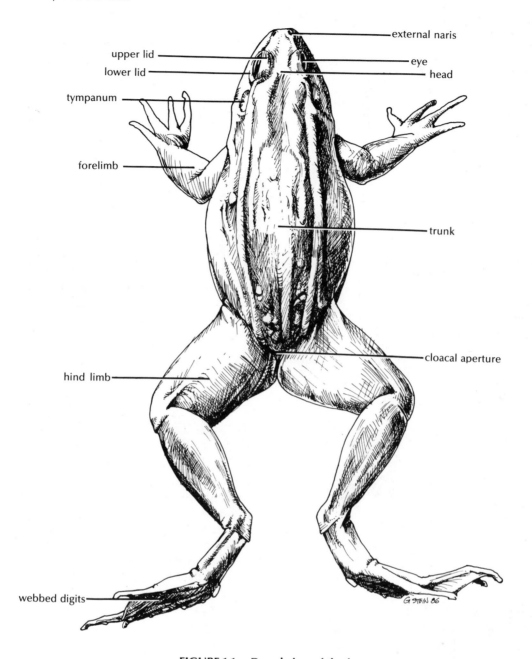

FIGURE 1.1. Dorsal view of the frog

predation through camouflage and secretion of poison, and it provides an important contribution as a respiratory structure by permitting air to diffuse across its outer layers to a rich capillary network beneath. It also actively absorbs water from the environment, making it unnecessary for the frog to "drink" water through its mouth.

The skin, or **integument**, of the frog consists of a loosely fit, smooth structure devoid of hair or scales. As in all land vertebrates, it consists of a superficial **epidermis** and a deep **dermis**. These layers may be observed by looking at a prepared slide of frog skin through a light microscope. The epidermis is composed of epithelium that is in a state of continual growth at the **basement layer**. The superficial layer of the epidermis consists of nonliving cells and is called the **cornified layer**. The epidermis is very thin relative to that of mammals. The dermis is composed of connective tissue that contains large numbers of mucus glands, poison glands, and chromatophores. The latter provide the skin with its coloration.

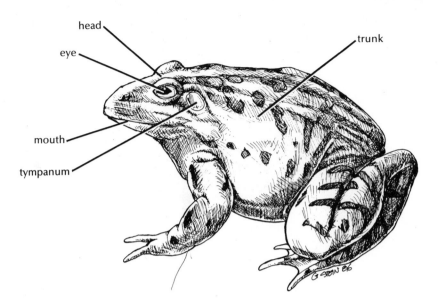

FIGURE 1.2.   **Lateral view of the frog**

The Skeletal System

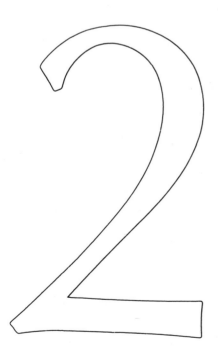

THE SKELETAL SYSTEM of the adult frog is primarily composed of bone, with joints (articulations) and some regions of the skull consisting of the more pliable tissue cartilage. In the larval frog (tadpole), the skeleton is entirely cartilaginous. The skeleton of the frog, like the skeleton of all backboned (vertebrate) animals, is an **endoskeleton**, for it lies within the soft tissue of the body. It plays several important roles, including **support** of soft body tissues, a site for muscle attachment to make **movement** possible, **protection** for vital organs such as the brain and spinal cord, and **storage** of calcium and phosphorus in the form of mineral salts.

For descriptive purposes, the skeleton may be divided into two divisions, the **axial skeleton**, which contains the bones that lie along the central vertical axis of the body, and the **appendicular skeleton**, which consists of bones that lie lateral to the central axis. The axial skeleton is composed of the bones of the skull, the vertebral column, and the sternum. The appendicular skeleton contains the bones of the pectoral girdle and forelimbs and the pelvic girdle and hind limbs. Study the complete (articulated) skeleton of the frog in Figure 2.1, and a mounted frog skeleton if one is provided by your instructor, to orient yourself with its general organization before proceeding.

BONES OF THE AXIAL SKELETON

Using the following illustrations and descriptions of bones as a guide, identify the bones of the axial skeleton and their major features on the articulated skeleton of the frog and disarticulated bones that may be available in your lab. The number of bones present in each group is given in parentheses following the name of each skeletal component.

SKULL (33)

The skull may be divided into two portions: the **cranium**, which encloses and protects the brain, and **facial bones**, which include most of the bones of the eye orbit, nose, cheek, and jaw. Notice that all of the bones of the skull are paired except the **parasphenoid**. Examine first

4

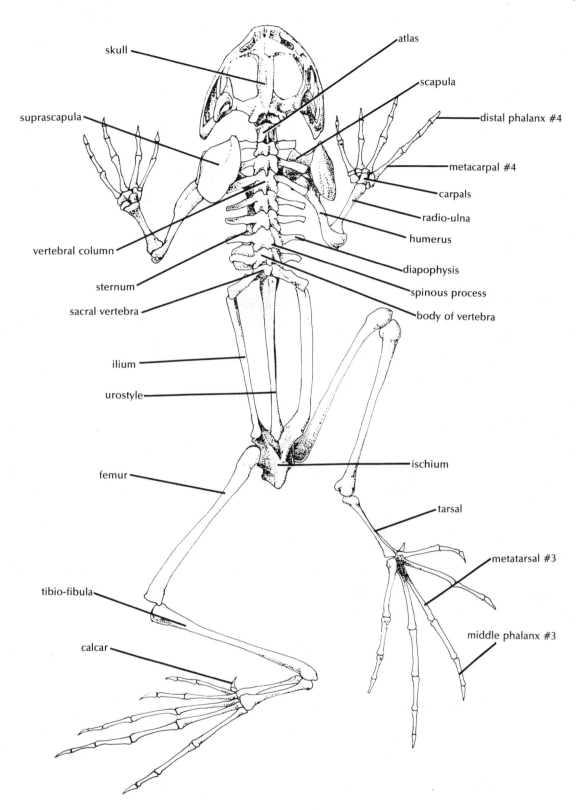

skull

atlas

suprascapula

scapula

distal phalanx #4

metacarpal #4

carpals

radio-ulna

humerus

vertebral column

diapophysis

sternum

spinous process

sacral vertebra

body of vertebra

ilium

urostyle

ischium

femur

tarsal

metatarsal #3

tibio-fibula

middle phalanx #3

calcar

FIGURE 2.1. Dorsal view of complete skeleton

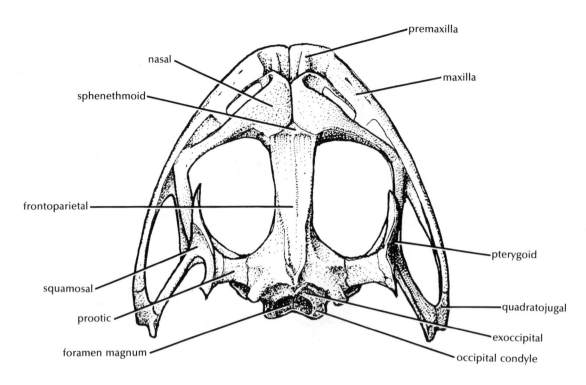

FIGURE 2.2.   Skull, dorsal view

the dorsal (Fig. 2.2) and ventral (Fig. 2.3) aspects of the skull, followed by the lateral aspect with the mandible included (Fig. 2.4).

Frontoparietal (2): flat bones that form the roof of the cranium.

Sphenethmoid (2): Located rostral to the frontoparietals and ventral to them, they form the anterior end of the cranium and enclose the olfactory tracts. Olfactory nerves pass through their small holes.

Nasal (2): Located rostral to the sphenethmoids on the dorsal side, they form a roof over the nasal cavity.

Exoccipital (2): Located caudal to the frontoparietals, they form the caudal portion of the cranium. Through the center of the exoccipitals passes the **foramen magnum**, a large opening that allows the spinal cord to enter the cranial cavity and unite with the brain. Also notice the rounded process on the caudal surface of each exoccipital. These are called **occipital condyles**, and they articulate with the first vertebra. The small holes at the base of the condyles allow passage of cranial nerves IX and X.

Prootic (2): Located on the dorsal side, they join with the caudal portion of the frontoparietals to form the caudal part of the eye orbit. They fuse with the exoccipitals along their caudal margins, forming a united bone called the **otoccipital**. The prootics also house the inner ear.

Squamosal (2): T-shaped bones that are visible on the dorsal and lateral aspects. They form the caudal border of the orbit.

Maxilla (2): paired bones that form most of the upper jaw. Notice the loosely attached row of teeth. The teeth are small, conical, and all the same type (**homodont**).

Premaxilla (2): paired bones that form the medial portion of the upper jaw. The premaxillae join together at the midline and connect to the maxillae laterally.

Quadratojugal (2): paired bones that form the caudal portion of the upper jaw. The quadratojugals connect the maxillae to the squamosals.

Parasphenoid (1): a dagger-shaped bone that is visible ventrally on the midline. It forms the floor of the cranial cavity.

Vomer (2): Located caudal to the premaxillae on the ventral side, they form the floor of the nasal cavity. Each bears several teeth on its caudal edge.

Palatine (2): Also visible ventrally, they bridge the maxillae to the sphenethmoids.

Pterygoid (2): three-pronged bones that are located medial to the maxillae on the ventral side.

Quadrate (2): small bones that form the articular surfaces between the upper and lower jaws.

Dentary (2): the middle portion of the lower jaw, or **mandible**.

Angular (2): the caudal portion of the mandible. At their

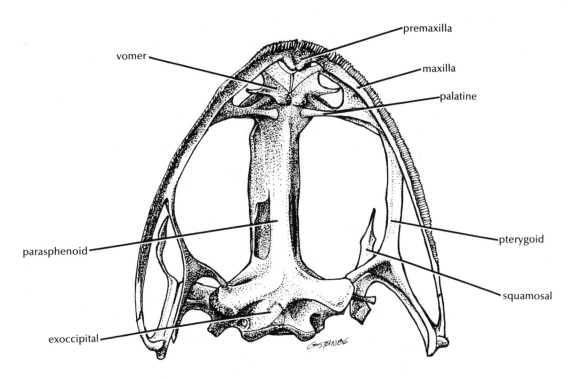

FIGURE 2.3.   Skull, ventral view, tilted laterally to the left side

caudal end is a projection called the **coronary process**. Each angular articulates with the quadrate bones to form the movable jaw.

Mentomeckelian (2): Also called the mentals, these small bones form the median connection between the right and left halves of the mandible and function in the action of tongue flipping.

VERTEBRAL COLUMN (10)

The vertebral column is composed of nine interlocking vertebrae and a highly modified bone called the **urostyle** at the caudal end. Notice that there are no ribs attached as there are in mammals. Identify the following features of the vertebral column (Figs. 2.1, 2.5).

Common features of vertebrae: The second through the eighth vertebrae share the following features:

Body: Also called the centrum, this is the central portion.

Neural arch: a major arch extending from the body and covering the neural canal dorsally.

Pedicle: a constricted portion of the neural arch near its union with the body.

Spinous process: a prominent dorsal projection.

Diapophyses:[1] Also called transverse processes, they are lateral projections.

Zygapophyses:[1] articular processes at the site of joints. There are a cranial pair and a caudal pair on each vertebra.

Neural canal: permits passage of the spinal cord.

Atlas (1): The atlas is the vertebra at the cranial end of the column which articulates with the skull at the occipital condyles. It is a modified structure in that it lacks diapophyses and cranial zygapophyses.

Sacral vertebra (1): the ninth vertebra; it is modified for articulation with the pelvic girdle as its diapophyses are enlarged and it lacks caudal zygapophyses.

Urostyle (1): a long, bladelike bone that extends caudally from the sacral vertebra. It represents a series of postsacral vertebrae that are fused together.

STERNUM (1)

The sternum serves as a point of attachment for certain muscles in the chest region. Because the frog lacks ribs, there is no true thoracic cavity as is found in reptiles, birds, and mammals. Beginning cranially, identify the following sections of the sternum (Figs. 2.1, 2.5):

Episternum: a thin circle of cartilage at the cranial end.

---

[1]The terms *diapophysis* and *zygapophysis* are actually taken from similar processes in the vertebrae of mammals. They are used here because better terms are not currently in use.

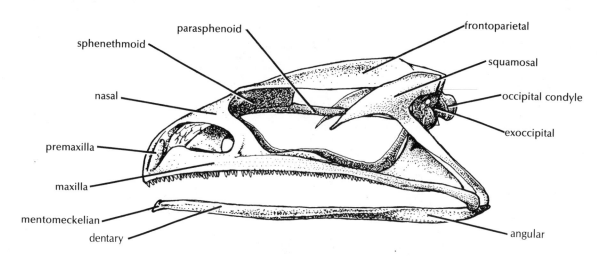

FIGURE 2.4. Skull, lateral view

Omosternum: a small section of bone caudal to the episternum.

Mesosternum: a rod-shaped bone.

Xiphisternum: a heart-shaped section composed of cartilage at the caudal end.

BONES OF THE APPENDICULAR SKELETON

The appendicular skeleton may be divided into four regions: the pectoral girdle, the cranial appendages or forelimbs, the pelvic girdle, and the caudal appendages or hind limbs.

PECTORAL GIRDLE (8)

The bones of the pectoral girdle provide attachment for the forelimbs to the body trunk. Identify the following (Figs. 2.1, 2.5):

Coracoid (2): The ventral portion of the pectoral girdle, the paired coracoids connect the sternum to the scapulae.

Clavicle (2): The cranial portion of the girdle, the paired clavicles provide a connection between the sternum and the scapulae.

Scapula (2): forms the lateral portion of each shoulder blade. The scapulae connect the clavicles and coracoids with the flat suprascapulae (below). At the junction of the scapula, clavicle, and coracoid on each side is a socket that is the point of articulation between the pectoral girdle and the forelimb, called the **glenoid cavity**.

Suprascapula (2): a broad, flat structure composed of bone and cartilage that forms the dorsal portion of the shoulder blade.

FORELIMB (42)

Identify the following bones of the forelimbs (Figs. 2.1, 2.5):

Humerus (2): the single bone of the upper forelimb, or **brachium**. The humerus contains a ridge, called the **deltoid ridge**, which serves as a point of attachment for the deltoid muscle.

Radio-ulna (2): the single bone of the lower forelimb, or **antebrachium**. It is formed by the fusion of the radius and ulna to provide increased strength.

Carpals (10): small bones of the wrist region.

Metacarpals (8): rod-shaped bones of the hand.

Phalanges (20): small bones that make up the four digits of each hand. Counting from medial to lateral, digits 1 and 2 each have two bones (proximal and distal), and digits 3 and 4 each have three (proximal, middle, and distal).

PELVIC GIRDLE (6)

The pelvic girdle provides a point of attachment for the bones of the hind limb to the axial division of the skeleton. The bones of the pelvic girdle are collectively referred to as **innominate bones**. Identify the following (Figs. 2.1, 2.5):

Ilium (2): long, narrow bones that unite the sacral vertebra to the hip joint.

Ischium (2): small bones that form the caudal portion of the pelvic girdle.

Pubis (2): forms the ventral portion of the pelvic girdle. The large socket that is formed by parts of the ilium, ischium, and pubis is called the **acetabulum**. It articulates with the head of the femur (below).

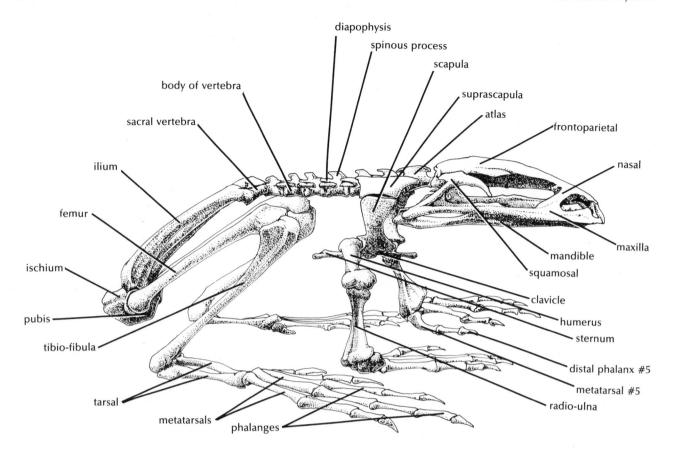

FIGURE 2.5. Lateral view of skeleton

HIND LIMB (46)

The bones of the hind limbs are specialized to permit the leaping activity of the frog. Identify the following (Figs. 2.1, 2.5):

Femur (2): The long bone in the cranial portion of the hind limb (thigh), it extends from its union with the acetabulum at the hip joint to the knee joint.

Tibio-fibula (2): a fused bone that articulates with the femur at the knee joint proximally and with the tarsals at the ankle joint distally.

Tarsal (4): two bones on each hind limb that form the elongated ankle and proximal portion of the foot. The medial tarsal is called the **tibiale**, and the lateral is the **fibulare**.

Metatarsal (10): the five bones of each foot that support the sole of the foot.

Phalanges (28): the small bones that form the five digits on each foot. Counting from medial to lateral, the first two digits each contain two bones (proximal and distal), the third and fifth digits each contain three bones (proximal, middle, and distal), and the fourth digit contains four (proximal, two middle, and distal). On the medial side of each foot is an additional toe, the **calcar**.

The Muscular System

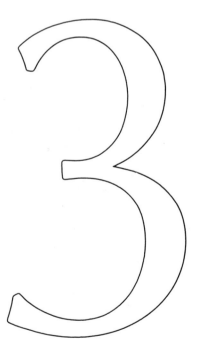

MUSCLE IS an important type of tissue that is found throughout the body of all vertebrates. There are three basic types of muscle tissue: **smooth** or **visceral muscle**, which forms part of the walls of visceral organs and blood vessels; **cardiac muscle**, which forms the bulk of the heart; and **skeletal** or **somatic muscle**, which lies deep to the skin and is attached to bones. Smooth and cardiac muscle are not considered to be components of the muscular system but are instead part of the organ and organ system that they help to form.

The muscular system consists only of the type of muscle that is attached to bones, the **skeletal muscle**, and its associated connective tissue. Skeletal muscle is composed of muscle cells, or **fibers**, that are arranged in parallel bundles to form the muscle unit. Surrounding individual fibers, groups of fibers called **fasciculi**, and entire muscles is an extensive network of connective tissue. This connective tissue, or **deep fascia**, continues out of the body of a muscle to form the means of attachment to a bone. A narrow band of deep fascia that extends from a muscle to a bone is called a **tendon**, and the broad, thin sheet of deep fascia that attaches a muscle to a bone is known as an **aponeurosis**.

During the contraction of a muscle, one end remains fairly stationary and the opposite end moves the bone to which it is attached as the muscle shortens. The more stable end is called the **origin** of the muscle, and the mobile end is known as the **insertion**. In the case of limb muscles, the origin is proximal and the insertion is distal.

The primary function of the muscular system is to provide skeletal movement. Movement is produced when a contracting muscle pulls an articulating bone toward a stationary bone. The various types of body movements (**actions**) resulting from skeletal muscle contractions are listed in Table 1.

In this chapter you will study the muscular system of the frog through dissection of preserved specimens. Each major muscle is listed according to its regional location and identified by a description of its point of origin, point of insertion, and primary action.

TABLE 1. Movements of Articulating Bones

Action	Description of Movement
Flexion	Decrease in angle between articulating bones
Extension	Increase in angle between articulating bones
Abduction	Movement away from body axis
Adduction	Movement toward body axis
Rotation	Movement around a central axis
Supination	Lateral rotation of the hand upward
Pronation	Lateral rotation of the hand downward
Eversion	Rotation of the sole of the foot outward
Inversion	Rotation of the sole of the foot inward
Circumduction	Flexion, extension, abduction, adduction, and rotation

BEGINNING YOUR DISSECTION

The primary goal of dissection is to bring into view structures that cannot readily be seen in their normal environment. As you will see, this goal is achieved by working from the outer surface of the specimen inward. Once the body structures are exposed, they may be examined and identified. Ideally, the complete dissection of your frog will permit you to study all the muscles and internal structures in their proper location in the body. As you proceed through the dissection protocol, keep the following points in mind:

1. *Follow the directions in this manual as they are presented in sequence,* much as you would in a cookbook. This should prevent you from getting lost or otherwise confused.
2. *Utilize care and patience.* Dissection is not merely "cutting up" an animal; it is a careful process of separating different parts from each other. Great care must be exercised at all times to avoid damaging structures before they have been identified. To keep unnecessary damage to a minimum, thoroughly examine the area you are about to cut into or pull apart before doing so. In other words, look at the bigger picture first.
3. *Use the right instrument for the job.* Correct tools are a necessity whenever precision is desired. Below is a list of common dissection instruments and a brief description of their proper uses:
   Blunt probe: a rigid 5-inch steel instrument with a blunt, bent tip. This is useful for gentle manipulation of muscles and internal organs.
   Scissors: usually 4–6 inches long. Scissors should be used to cut through skin, muscles, and other large structures.
   Scalpel: usually 5 inches long with replaceable blades. The scalpel should be used to make small incisions.
   Needle probe: a 3-inch needle attached to a wooden handle. This may be used as a pointer or to attach the specimen to the dissecting tray.
   Forceps: about 5 inches long. These are commonly called "tweezers." They are used to grasp small objects.

SKINNING THE FROG

Your frog specimen was sent to you in a preserved state. As it appears before you, it may be saturated with fluid or dry. This difference in appearance is due to the different methods used by biological supply companies to prepare the specimen after they have been preserved. In either case, your frog was preserved by saturating it after death in a dilute formalin solution. This chemical can be damaging to clothing and irritating to exposed skin and eyes, so you should always wear plastic gloves and protective clothing when dissecting. A surgical face mask may be required for sensitive individuals.

The following skinning procedure is the first step in dissection. Follow the protocol outlined below when you are ready to begin:

1. Place your frog on its back in the dissecting tray. With your scissors, make a shallow cut through the skin just below the jawline. Continue this cut midventrally from the jaw to the cloacal opening. As you cut through the skin, pull upward and away from the underlying muscle to assure that you are not cutting too deeply.
2. From this cut make additional incisions through the skin around the cloacal opening, down the lateral surface of each limb, around the ankles and wrists, and around the head. In the head region, cut around the eyes and tympanum without damaging them.
3. With your fingers, pull the skin away from the underlying muscles until it is completely removed. Observe that the skin does not adhere to the muscles but is attached by connective tissue bridges called **septa**. This arrangement results in the presence of a potential space between the skin and muscles. In life this space is occupied by a colorless fluid called **lymph**. Also observe that there are no fat deposits between the skin and muscle, as there are in mammals. What significance do you think this has for a frog in cold weather?

MUSCLE DISSECTION

Muscle dissection involves the careful separation of muscles from each other. This is possible because the fibers of each individual muscle run parallel to each other between attachments. Individual muscles are visible and therefore separable when adjacent muscles contain fibers that travel in a different direction.

Examine the muscles on your frog closely. Locate the cleavage lines that separate adjacent muscles while comparing your specimen with the diagrams (Figs. 3.1, 3.2, and 3.3). If the lines are not visible, gently pull the muscles apart with your fingers until the natural areas of separation appear. Using a blunt probe to break the surrounding connective tissue, separate the muscles at the cleavage line by inserting the probe between adjacent muscles and working the instrument forward. Continue this until the probe tip resurfaces at the opposite cleavage line of each muscle.

You should follow each muscle to its points of origin and insertion as far as is practical. This is done by inserting the probe through both cleavage lines bordering a muscle and sliding the probe laterally along the muscle's length while pulling the muscle slightly outward. If this step is done with care, the probe will break through the connective tissue between adjacent muscles without causing any damage. Proceed to separate all muscles on one side of the body in this manner in order of their sequence, as presented below.

DORSAL HEAD AND SHOULDER MUSCLES

With your frog lying on its ventral side in the dissecting tray, identify and separate the following head and shoulder muscles on the dorsal side (Figs. 3.1, 3.2):

Temporalis: a small muscle caudal to the eye on the head. It originates from the squamosal and pterygoid bones of the skull and inserts at the mandible. Action: elevates the mandible.

Masseter: a muscle on the side of the head that extends from its origin at the squamosal bone to the mandible. Action: elevates the mandible.

Depressor mandibulae: located caudal to the tympanum. Its origin is at the frontoparietal bone, and its insertion is at the mandible. Action: depresses the mandible.

Dorsalis scapulae: a shoulder muscle caudal to the depressor mandibulae. Its origin is at the scapula, and its insertion is at the humerus of the forelimb. Action: extends the forelimb.

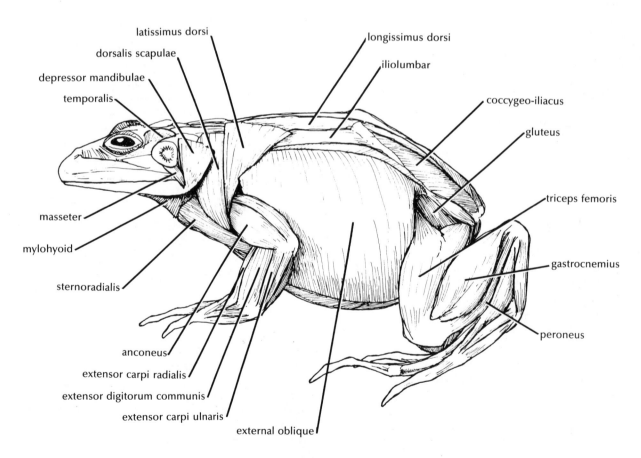

FIGURE 3.1. Superficial muscles, lateral view

Deltoid: located on top of the shoulder joint. It originates from the episternum, clavicle, and sternum and inserts on the humerus. Action: abducts the forearm.

VENTRAL HEAD AND SHOULDER MUSCLES

Turn your specimen over on its back to expose its ventral side. Identify and separate the following muscles (Fig. 3.3):

Mylohyoid: a large, thin sheet of muscle that covers the throat. Its origin is at the rostral end of the mandible, and its insertion is at the center line of the mandible. Action: raises the floor of the mouth. A number of muscles are located deeper than the mylohyoid and are mainly involved in moving the tongue and throat.

Sternoradialis: located caudal to the mylohyoid in the shoulder region and across the cranial chest region. It is sometimes called the **coracoradialis**. Its origin is at the sternum, and its insertion is at the radio-ulna. Action: draws the arm forward.

VENTRAL MUSCLES OF THE TRUNK

Continue the dissection with your frog on its dorsal side. Refer often to Figure 3.3.

Pectoralis major: a large muscle of the chest region that has three points of origin: at the medial coracoid, at the cranial coracoid, and at the lateral aponeurosis of an adjacent muscle called the **rectus abdominis**. All three portions insert at the humerus. Action: adducts, flexes, and medially rotates the forelimb.

Cutaneous pectoris: a large muscle covering the cranial abdomen. Its origin is at the xiphisternum and the rectus abdominis aponeurosis. It inserts onto the underside of the skin near the sternum. Action: tenses skin of the cranial abdomen.

Rectus abdominis: a large muscle that extends along the ventral midline from the pubic symphysis to the xiphisternum. Its origin and insertion are dependent upon which end of the muscle is fixed during a given action and are therefore reversible. Action: flexes the vertebral column.

DORSAL MUSCLES OF THE TRUNK

Turn your specimen over onto its belly side in the dissecting tray to expose the dorsal side. Identify and separate the following trunk muscles (Figs. 3.1, 3.2):

Latissimus dorsi: located caudal to the dorsalis scapulae. Its origin is at the cranial end of the large sheet of fascia in the center of the back, called the **dorsal aponeurosis**. It extends from this fascia to insert at the humerus. Action: rotates the forelimb.

Longissimus dorsi: located caudal and medial to the latis-simus dorsi. This muscle originates from the urostyle and passes cranially to insert at the base of the skull. Its caudal half is located deep to the dorsal aponeurosis, and part of its cranial half passes deep to the latissimus dorsi. Action: extends the trunk and elevates the head.

Iliolumbar: located lateral to the longissimus dorsi. This smaller muscle has its origin at the sacrum and ilium and its insertion along the diapophyses of the middle vertebrae. Action: flexes and extends the vertebral column.

Coccygeo-iliacus: a caudal back muscle that extends from its origin at the urostyle to its insertion at the ilium. Action: supports the articulation of the coccyx with the ilium.

Cutaneous abdominis: a superficial muscle that connects the fascia of the skin at the ilium to the pubic symphysis. Action: tenses the skin of the back.

External oblique: a large sheet of muscle that extends from its origins at the scapula and the spinous processes of vertebrae to its insertions at the xiphisternum and the ventral midline, or **linea alba**. Action: constricts the abdomen.

Internal oblique: Make a shallow longitudinal incision along the lateral surface of the abdomen, taking special care to cut through only the most superficial muscle layer, the external oblique. The sheet of muscle lying deep to the external oblique is the internal oblique. Its origins are at the ilium and vertebrae, and its insertions are at the xiphisternum and the linea alba. Note that its fibers lie at right angles to the fibers of the external oblique. Actions: aids in respiration and aids the external oblique in constriction of the abdomen.

DORSAL MUSCLES OF THE FORELIMB

With your specimen on its belly, identify the following forelimb muscles (Figs. 3.1, 3.2):

Anconeus: In higher vertebrates this muscle is usually called the **triceps brachii**. Its origin is at the humerus, and its insertion is at the ulna side of the fused radio-ulna. Action: extends the forearm at the elbow.

Extensor digitorum communis: the largest muscle in the forearm. It originates at the humerus and inserts on the surface of digits 2, 4, and 5. Action: extends the digits and the hands.

Extensor carpi ulnaris: located medial to the extensor digitorum communis. Its origin is at the humerus, and its insertion is at the dorsal surface of the wrist and digits. Action: extends the hand and digits.

Extensor carpi radialis: located lateral to the extensor digitorum communis. Its origin is at the humerus, and its insertion is at the dorsal surface of the wrist and digits. Action: extends the hand and digits.

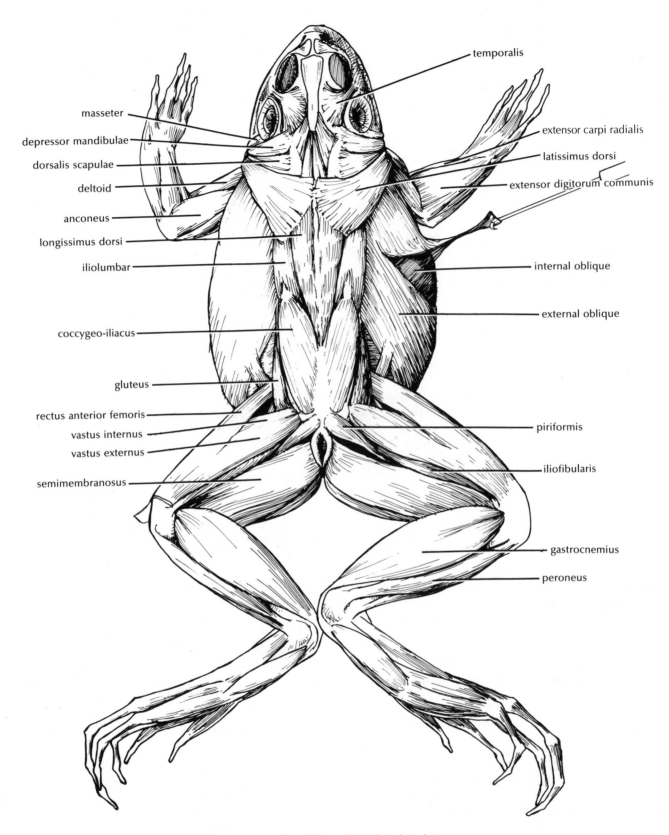

temporalis

masseter

depressor mandibulae

dorsalis scapulae

deltoid

anconeus

longissimus dorsi

iliolumbar

coccygeo-iliacus

gluteus

rectus anterior femoris

vastus internus

vastus externus

semimembranosus

extensor carpi radialis

latissimus dorsi

extensor digitorum communis

internal oblique

external oblique

piriformis

iliofibularis

gastrocnemius

peroneus

FIGURE 3.2. Superficial muscles, dorsal view

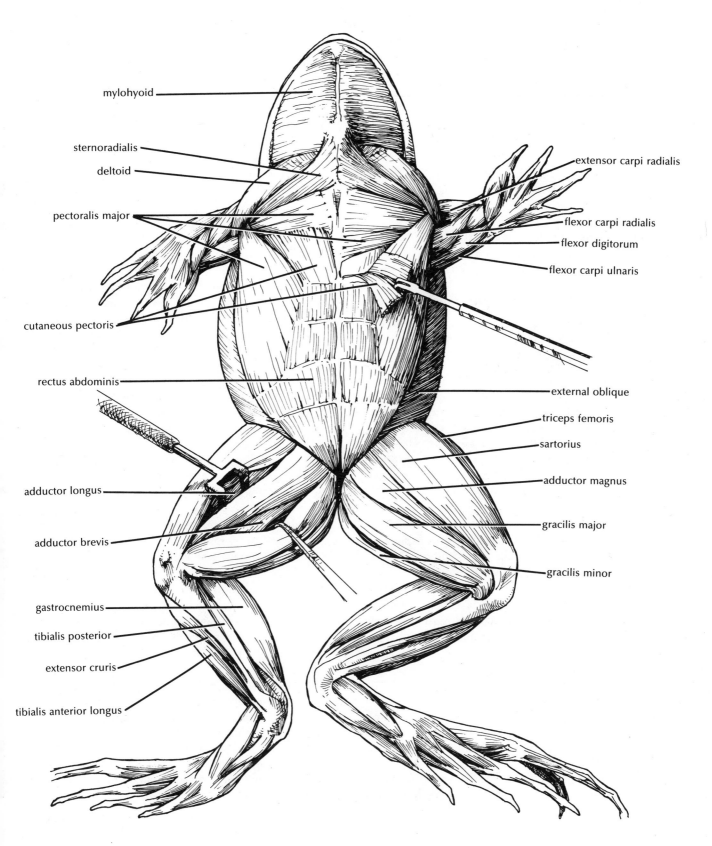

mylohyoid

sternoradialis

deltoid

pectoralis major

cutaneous pectoris

rectus abdominis

adductor longus

adductor brevis

gastrocnemius

tibialis posterior

extensor cruris

tibialis anterior longus

extensor carpi radialis

flexor carpi radialis

flexor digitorum

flexor carpi ulnaris

external oblique

triceps femoris

sartorius

adductor magnus

gracilis major

gracilis minor

FIGURE 3.3.   Superficial muscles, ventral view

VENTRAL MUSCLES OF THE FORELIMB

Turn your specimen over onto its back in the dissecting tray. With the ventral side once more exposed, identify the following forelimb muscles (Fig. 3.3):

Flexor digitorum: This muscle in the midventral forearm is also called the **palmaris longus**. Its origin is at the humerus, and its insertion is at the fascia of the digits located in the palm. Action: flexes the digits and the hand.

Flexor carpi ulnaris: located medial to the flexor digitorum. Its origin is at the humerus, and its insertion is at the ventral surface of the wrist and digits. Action: flexes the wrist and hand.

Flexor carpi radialis: located lateral to the flexor digitorum. It also originates at the humerus and inserts at the wrist and digits. Action: flexes the wrist and hand.

VENTRAL MUSCLES OF THE HIND LIMB

Identify the following muscles of the hind limb with your specimen remaining on its back side in the dissecting tray (Fig. 3.3):

Sartorius: a large, flat midventral muscle. Its origin is at the pubis, and its insertion is at the tibio-fibula. Action: at the hip joint it flexes the thigh; at the knee joint it flexes the shank.

Gracilis major: a large, flat muscle located medial to the sartorius. Its origin is at the pubic symphysis, and its insertion is at the proximal end of the tibio-fibula. Action: at the hip joint it extends the thigh; at the knee joint it flexes the shank.

Gracilis minor: a thin muscle immediately medial to the gracilis major. This muscle, which is not separate in most higher vertebrates, shares the same origin, insertion, and action as the gracilis major.

Adductor magnus: This muscle may be seen only partially as a portion of it lies deep to the sartorius. Its origin is at the ischium and pubis, and its insertion is at the femur. Action: adducts the thigh at the hip joint.

Adductor longus: A portion of this muscle may be observed lateral to the sartorius. Its origin is at the pubis, and it inserts at the femur. Action: it also adducts the thigh.

Adductor brevis: This is a small muscle that may be observed by pulling the sartorius at its origin laterally. It originates at the pubis and inserts at the femur. Action: adducts the thigh.

Tibialis anterior longus: a small muscle of the shank that is anterior to the tibio-fibula. Its origin is at the femur, and its insertion is at the two tarsal bones. Action: extends the foot at the ankle.

Tibialis posterior: a small muscle located posterior to the tibio-fibula. It originates at the tibio-fibula and passes down the shank to insert at the tarsal bones. Action: flexes the foot at the ankle.

Extensor cruris: a small muscle located cranial and partially deep to the tibialis anterior longus. Its origin is at the femur, and its insertion is at the tibio-fibula. Action: extends the shank.

DORSAL MUSCLES OF THE HIND LIMB

Turn your specimen over on its belly side to reveal the dorsal muscles once again. Identify the following hind limb muscles (Figs. 3.1, 3.2):

Triceps femoris: a large muscle that consists of three portions, or heads, due to three different points of origin and a common insertion. They are : the **middle head**, or **rectus anterior femoris**, which originates at the cranial ilium; the **posterior head**, or **vastus externus**, which originates at the caudal ilium; and the **anterior head**, or **vastus internus**, which has its origin at the acetabulum. All three insert at fascia that is attached to the tibio-fibula. Action: all heads flex the thigh at the hip joint and extend the shank at the knee.

Gluteus: also called the **iliacus externus**, it is located medial to the triceps femoris. Its origin is at the ilium, and its insertion is at the femur. Action: rotates the thigh in a lateral direction.

Semimembranosus: a large muscle medial to the triceps femoris. It originates at the ischium and pubis and inserts at the tibio-fibula. Action: extends the thigh at the hip joint and flexes the shank at the knee.

Iliofibularis: a small muscle located between the triceps femoris and the semimembranosus. Its origin is at the ilium, and its insertion is at the tibio-fibula and femur. In most higher vertebrates it is called the **biceps femoris**. Action: extends and adducts the thigh and flexes the shank.

Piriformis: a small muscle near the cloacal orifice. Its origin is at the urostyle, and its insertion is at the femur. Action: extends and rotates the thigh in a lateral direction.

Gastrocnemius: the large muscle of the shank. It originates at the femur and passes down the shank to insert at the large **Achilles' tendon**. Action: flexes the shank at the knee and the foot at the ankle.

Peroneus: located lateral to the gastrocnemius on the shank. Its origin is at the femur, and its insertion is at the caudal extremity of the tibio-fibula. Action: abducts the shank and extends the foot.

Internal Anatomy

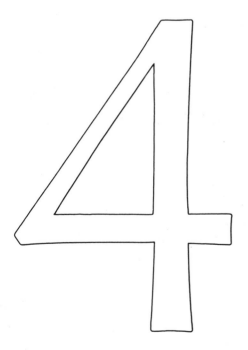

THE INTERNAL ORGANIZATION of the frog is representative of adult members of the Class Amphibia. Similar to all vertebrates, the frog has a large central cavity called the **coelom**, which is an epithelium-lined space that contains the visceral organs and a small amount of fluid. It is divided into two parts: a cranial portion around the heart called the **pericardial cavity** and the remainder called the **pleuroperitoneal cavity**. The two are partially separated by a fold called the **transverse septum**, which extends in an oblique direction. This septum is also present in fishes but travels vertically to separate the pericardial and pleuroperitoneal cavities. In certain reptiles, the cranial portion of the pleuroperitoneal cavity lies dorsally to the pericardial cavity to surround each lung. In these more advanced reptiles and in mammals, these cavities are called **pleural cavities** and permit increased efficiency in the breathing process. In mammals, the two parts of the coelom are completely separated by the muscular **diaphragm**. The diaphragm is in part embryonically developed from the primitive transverse septum. This embryonic origin of the diaphragm in mammals provides us with a clue to its evolutionary origin.

In this chapter, the internal anatomy of the frog is presented as it appears when you first expose the coelom and before manipulation of any body structures. This, then, represents an overview of the locations of the major internal organs of the frog that can be seen easily. In subsequent chapters, their categorization into body systems and their functional roles will be discussed.

DISSECTION OF THE BODY WALL

In order to expose the coelom, the body wall must first be dissected. To do this, follow the procedure below.

1. Place your frog on its back to expose the ventral side. Using your scissors, cut the body wall slightly to one side of the midventral line and extend the cut from the cloacal opening cranially to the pectoral girdle. Do not insert your scissors too deeply, or the internal organs you are about to study will be damaged. As a precaution when cutting, exert a slight upward pressure with the point of your scissors to lift the body wall from underlying structures.

2. Lift the cut edges of the abdominal wall. Note the large vein attached to the body wall that extends along the ventral midline. This is called the **ventral abdominal vein**. Free this vein by cutting the connective tissue that attaches it to the body wall and leave the vein intact.

3. Make a transverse cut in the body wall behind the pectoral girdle and extend it from shoulder to shoulder. Make a similar cut in the pelvic region. Now you may pin back or cut off the portions of the body wall that had been freed to expose the coelom.

MAJOR ORGANS OF THE COELOM

Closely inspect your frog's internal anatomy. Notice the large amount of space around the viscera. This space is the pleuroperitoneal cavity, and in life it contains a fluid called **lymph** that sluggishly circulates throughout the body. In mammals, lymph is carried throughout the body by way of a closed series of vessels called lymphatic vessels.

Identify the following organs in your specimen with as little manipulation as possible. Use Figure 4.1 as a guide. Compare your specimen with others in the class and note the variations in organ size and shape that exist among individuals.

Lung: paired organs that are visible in the cranial extremity of the pleuroperitoneal cavity. They lie dorsal to the liver.

Heart: located midventral and cranial. It is enclosed within a membranous sac called the **pericardial sac**.

Liver: a large, dark brown organ that is the most conspicuous structure visible.

Stomach: Lift the left segment of the liver slightly to view the stomach beneath.

Small intestine: a long, coiled tube that is normally caudal and partially deep to the liver.

Large intestine: an enlarged caudal continuation of the small intestine. In some specimens it may be found deep to the small intestinal loops.

Urinary bladder: a small sac on the caudal floor of the pleuroperitoneal cavity.

Ovary: If your frog is a female, the ovaries will be visible as gelatinous bodies in the caudal region of the pleuroperitoneal cavity; they are partially obstructed from view by other organs. In some females, the ovaries may be the largest structures in the coelom if the animals were collected during the spring mating season.

Gallbladder: visible as a small swelling near the caudal edge of the right liver lobe.

Kidney: paired, reddish organs. In order to view the left kidney, carefully lift the stomach and small intestine and push them to the right side of the body.

Fat bodies: yellow clusters of fat attached to the cranial end of the kidney. They provide an important storage for food reserves. In females, the energy is channeled into egg production.

Testis: If your specimen is a male, the left testis may be observed as a yellow, bean-shaped organ located medial to the kidney.

Spleen: With the stomach and small intestine reflected to the right, the spleen may be viewed medial to the kidney and testis.

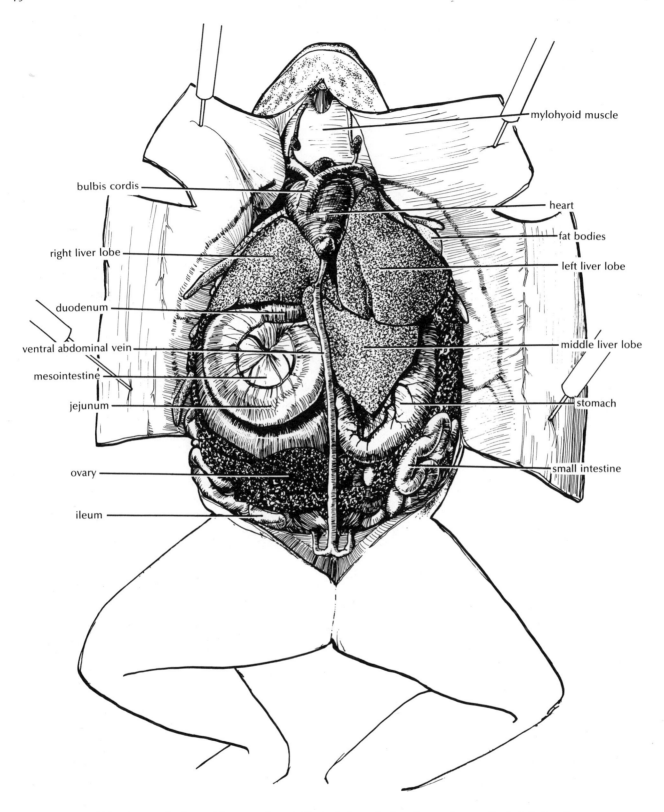

bulbis cordis

right liver lobe

duodenum

ventral abdominal vein

mesointestine

jejunum

ovary

ileum

mylohyoid muscle

heart

fat bodies

left liver lobe

middle liver lobe

stomach

small intestine

FIGURE 4.1.   Coelom with organs intact. Ventral view of a female specimen collected during the spring mating season (note the swollen ovaries).

The Digestive System

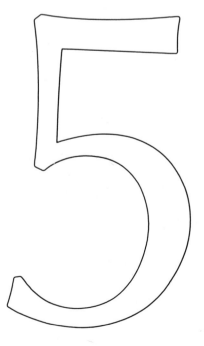

THE DIGESTIVE SYSTEM is one long, continuous, compartmented tube that extends through the pleuroperitoneal cavity from the mouth to the anus. Known as the **gastrointestinal (G.I.) tract** or **alimentary canal**, it is here that the functions of mechanical digestion, chemical digestion, absorption of nutrients, and storage and elimination of solid waste material take place. Each of these processes occurs within a compartment of the tract that is specialized to accommodate it. In the tadpole, these compartments are able to process plant and animal sources of food, but in the adult frog the compartments are adapted for the conversion of animal food only. The specialized compartments or organs of the G.I. tract are the mouth, pharynx, esophagus, stomach, small intestine, large intestine, and cloaca.

Also included in the digestion process are a number of structures that are closely associated with the G.I. tract either because they are located within it or because they communicate with it by means of a duct. These structures, called **accessory organs**, include the teeth, tongue, liver, gallbladder, and pancreas.

ORGANS OF THE DIGESTIVE SYSTEM

The organs and associated structures of the digestive system will be discussed sequentially from the mouth to the cloaca. Follow the dissection protocol below and refer often to Figures 5.1 through 5.5 as you proceed.

MOUTH

In order to view the mouth and its contents, open it as far as possible. If necessary, cut the corners of the jaws with bone clippers to obtain adequate exposure. Using Figures 5.1 and 5.2 as guides, identify the following structures of the mouth, or oral cavity:

Maxillary teeth: small, sharp teeth of equal size and shape.

Vomerine teeth: a pair of tooth patches partially hidden by mucus membrane on the roof of the mouth. They act with the tongue to hold onto food, preventing live prey from escape.

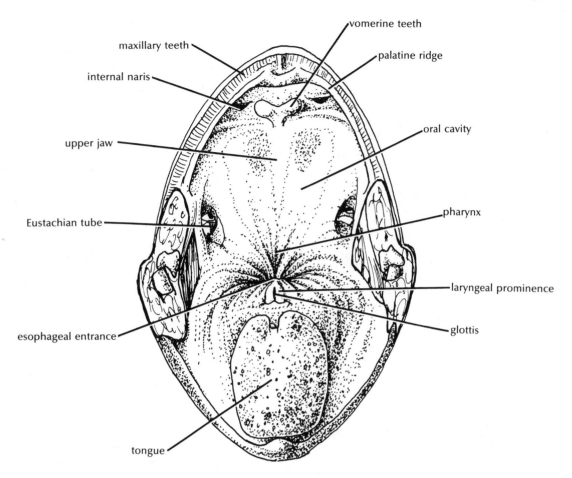

FIGURE 5.1.   Oral cavity and pharynx with jaws disarticulated. The corners of the mouth have been cut to permit greater visibility.

Palatine ridge: a fold of mucus membrane caudal to the maxillary teeth on the roof of the mouth.

Internal naris: a pair of holes that open into the nasal cavity. To demonstrate this, pass a thin probe through an external naris and notice the point at which it enters the oral cavity.

Tongue: forms part of the floor of the mouth. Pull the tongue forward and notice its attachment to the edge of the lower jaw. The tongue of the frog contains muscles that permit it to protrude and retract rapidly. To catch an insect, the frog flips out the tongue, extends it, and wraps it around the insect.

PHARYNX

The pharynx is a chamber immediately caudal to the mouth. It receives food from the mouth and air from the internal nares. Notice the two large openings into the pharynx. These are the paired **Eustachian tubes**, which pass to the middle ear on each side. You can demonstrate this by poking a probe through the tympanum on one side and observing where the tip emerges.

ESOPHAGUS

The esophagus is a short tube that extends between the pharynx and the stomach. This may be observed by inserting your probe into the esophagus from its entrance at the pharynx. The esophagus moves food toward the stomach by muscular peristaltic contractions and by the beating of cilia that line its internal surface.

LIVER

The liver is the large, dark brown organ in the pleuroperitoneal cavity. It is composed of several sections or **lobes**: the **right lobe**, the **middle lobe**, and the **left lobe**. The left lobe is again divided into anterior and posterior sections. Some of the notable functions of the liver include the storage of fats and glycogen, the formation of urea (a waste product of protein metabolism), the destruction of toxic substances including dead or dying red blood cells, and the production of bile that is used for the digestion of fats within the small intestine. Because the liver plays an important role in fat storage, it increases in

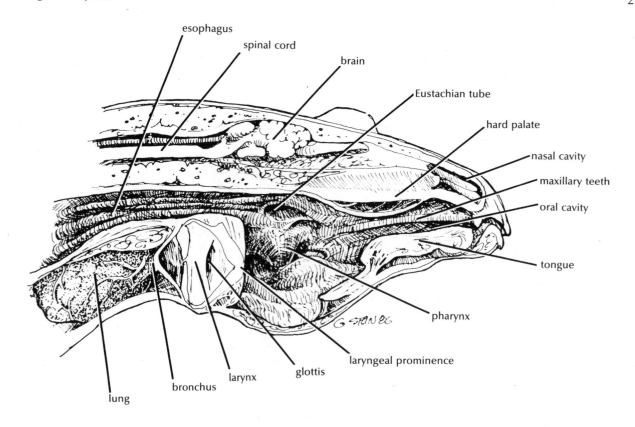

FIGURE 5.2.   Head region, lateral view of sagittal section

size (hypertrophies) during the fall season when the frog is preparing for dormancy.

GALLBLADDER

The gallbladder is a small, greenish sac deep to the right and middle lobes of the liver. To view it clearly, lift the liver lobes upward. The gallbladder provides a temporary storage site for bile as it passes from the liver. During digestion, bile passes into the small intestine by way of a small tube called the **common bile duct**.

STOMACH

The stomach appears as a J-shaped enlargement of the G.I. tract. To observe its entire shape, push the liver upward once again. With a sharp scalpel, make an incision through the stomach wall and observe the thick muscular layer that accounts for most of its thickness. This muscle is responsible for the mechanical digestion of food. Also observe the inner surface of the stomach. This is called the **mucosa**. Its numerous wrinkles and folds enable the stomach to expand dramatically.

SMALL INTESTINE

With the liver pushed upward, the small intestine can be observed as a long, coiled tube that extends from the stomach to the large intestine. It is divided into three segments that are continuous with each other: the **duo-denum**, which emerges from the caudal extremity of the stomach, the middle **jejunum**, and the **ileum**, which unites with the large intestine. Now with your scalpel, make an incision along a segment of the small intestine. Notice that the wall of the small intestine is thinner than that of the stomach. The inner surface contains numerous small projections called **villi**. The villi increase the absorptive surface area of the small intestine. The small intestine is highly adapted for the absorption of nutrients, which is its primary function.

PANCREAS

The pancreas is a small organ that is partially concealed by the peritoneal membrane (see below) deep to the middle lobe of the liver. It secretes a number of enzymes that chemically digest food within the small intestine. Its enzymatic secretions, commonly called **pancreatic juice**, pass into the common bile duct from the pancreas along with bile from the liver and gallbladder to empty into the duodenum. It also secretes the hormones, insulin and glucagon, which pass directly into the bloodstream to control sugar levels in the blood.

LARGE INTESTINE

The large intestine may be observed as an enlargement of the G.I. tract in the pelvic region. It extends between the ileum and the cloaca. Its opening into the

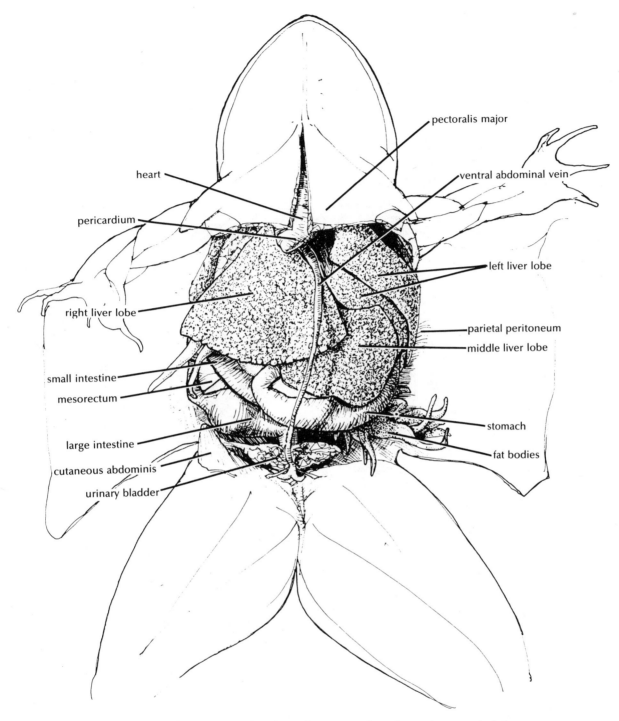

FIGURE 5.3.   Pleuroperitoneal cavity and contents of a male specimen, ventral view

cloaca is called the **anus**. The large intestine absorbs water from waste that arrives from the small intestine. This activity results in the formation of solid waste, called **feces**.

CLOACA

The cloaca is a tube that extends from its union with the large intestine to the exterior. The cloaca may be ob-

served by pushing aside the urinary bladder at the floor of the pleuroperitoneal cavity. It is a common chamber for the digestive, urinary, and reproductive systems. Its opening to the exterior is called the **cloacal aperture**.

PERITONEUM

The peritoneum is not an organ but an extensive membrane that lines the pleuroperitoneal cavity and covers

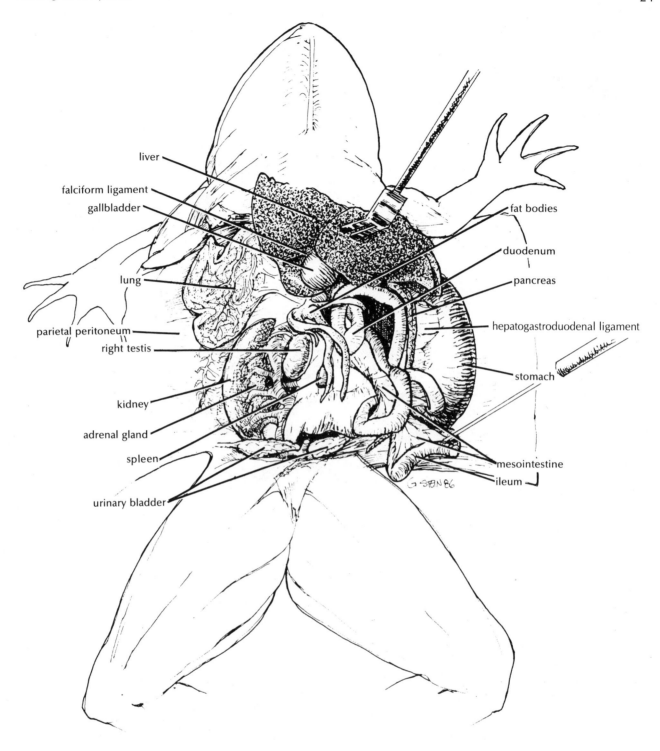

FIGURE 5.4.   Pleuroperitoneal cavity, ventral view. The liver in this male is reflected cranially and the G.I. tract is pulled laterally to the left side. The ventral abdominal vein has been cut and removed.

the visceral organs. The portion of the peritoneum that lines the internal surface of the body wall is called the **parietal peritoneum**, and the portion that covers most of

the visceral organs is the **visceral peritoneum**.

In addition to the peritoneal membranes, there are **peritoneal folds**. These extensions of the peritoneum

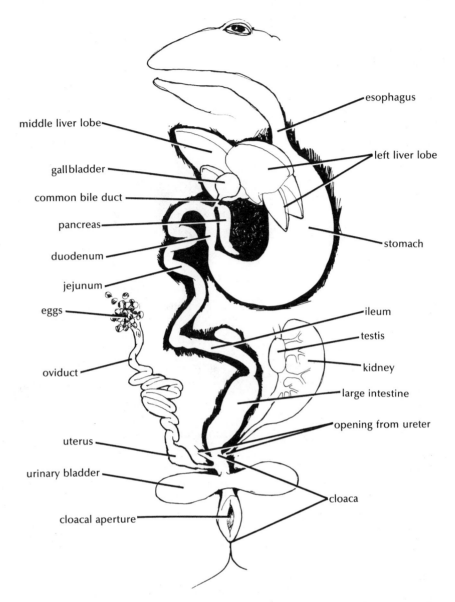

FIGURE 5.5.   Schematic diagram of G.I. tract organs. Female urogenital organs are shown on the right side and male organs on the left side to illustrate their relationship to the cloaca.

help suspend the visceral organs within the pleuroperitoneal cavity. Locate the following peritoneal folds in your frog (Figs. 5.3, 5.4):

Dorsal mesentery: a double layer of peritoneum that anchors visceral organs to the dorsal body wall. It consists of the following:

> **Mesogaster**: attaches to the stomach.
> **Mesointestine**: attaches to the intestine.
> **Mesorectum**: attaches to the rectum, or caudal portion of the large intestine.

Ventral mesentery: a reduced portion of the peritoneum that persists in adult frogs as several specialized ligaments. These provide support for some visceral organs to the ventral body wall and include the following:

> **Falciform ligament**: attaches to the cranial end of the liver but is very difficult to locate.
> **Hepatogastroduodenal ligament**: connects the stomach to the liver and duodenum.

Lateral mesentery: folds of peritoneum that support the gonads of both sexes to the lateral and dorsal body walls. They include the following:

> **Mesovarium**: supports the ovaries of the female.
> **Mesorchium**: supports the testes of the male.

The Respiratory System

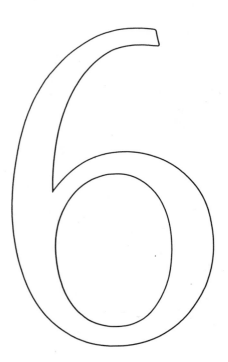

THE PRIMARY FUNCTIONS of the respiratory system are to provide the body with a continuous supply of oxygen and to remove the metabolic waste product, carbon dioxide. In fish and tadpoles, this is accomplished by the diffusion of these gases between the water and specialized structures called **gills**. Diffusion occurs rapidly because the gills are directly exposed to the water medium. In adult frogs, the gills are lost during metamorphosis, and their function is replaced by three regions of the frog's body: the skin (**cutaneous respiration**), an internal pair of lungs (**pulmonary respiration**), and to a lesser extent, the mucus membrane of the mouth and pharynx (**buccopharyngeal respiration**). In each of these regions air is separated from the bloodstream only by a thin, moist membrane—precisely the condition required for aeration of the blood.

In reptiles, birds, and mammals, pulmonary respiration is the only means of air exchange. Air is drawn into the lungs by expansion of the pleural cavity containing the lungs through muscular contraction of the diaphragm and rib muscles (intercostals). However, the frog lacks movable ribs and must therefore force air into the lungs with its mouth. This is accomplished by rhythmical contractions of the floor of the mouth, which forces air in through the nares. Air is pushed into the lungs from the mouth by closing the nares and raising the floor of the mouth. Expiration is performed by contraction of the abdominal muscles, which pushes air out of the elastic lungs. By itself, this method of respiration is not an efficient process. The frog is able to aerate its blood enough to survive only by combining pulmonary respiration with the equally important process of cutaneous respiration, with minimal contributions from buccopharyngeal respiration.

In this chapter, the organs and associated structures involved in pulmonary and buccopharyngeal respiration will be studied. The skin was previously discussed in Chapter 1 (p. 2). These include the nasal cavity, mouth, pharynx, larynx, bronchi, and lungs.

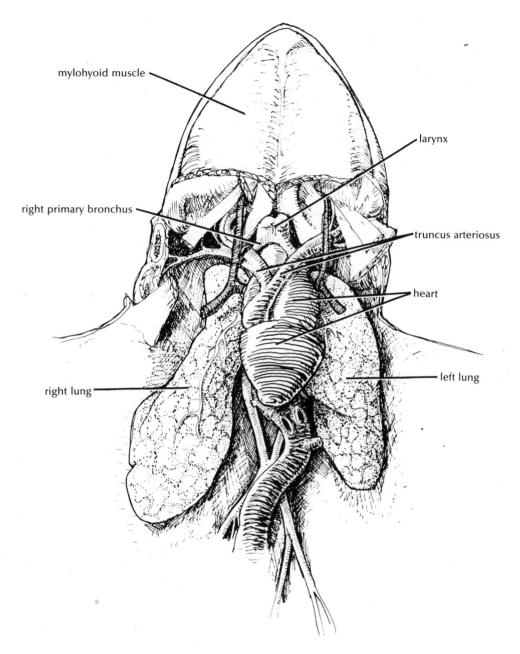

mylohyoid muscle

larynx

right primary bronchus

truncus arteriosus

heart

right lung

left lung

FIGURE 6.1.   Ventral view of head and cranial trunk region, with stomach and liver removed

ORGANS OF THE RESPIRATORY SYSTEM

The respiratory organs involved in pulmonary and buccopharyngeal respiration in the frog will be discussed in sequence, consistent with the pathway of air as it travels toward the lungs.

NASAL CAVITY

The nasal cavity of the frog cannot be dissected properly because of its small size. Examine instead the location of the **external nares**, the holes where air enters the nasal cavity, and the **internal nares**, the holes where air exits the nasal cavity and enters the mouth.

MOUTH

As discussed in Chapter 5 (p. 20), the mouth receives the internal nares. With the mouth held wide open, notice the pair of small openings lateral to the base of the tongue if your frog is a male. These holes open into **vocal sacs** located under the submaxillary muscle deep to the tongue. They serve as resonating chambers for mating-season calls.

PHARYNX

With the mouth remaining open, examine once again the pharynx located immediately caudal to the mouth. At the caudal end of the pharynx are two openings: the previously discussed esophagus that passes to the stomach, and a slitlike opening ventral to it called the **glottis**. The glottis opens into the larynx (discussed below). Surrounding the glottis is a swelling called the **laryngeal prominence**, which is composed of cartilage overlain with mucus membrane.

LARYNX

The larynx is a cartilaginous tube that extends from the glottis, which opens into it, to its division into two bronchi. The bronchi pass to each lung. The larynx contains the vocal cords, which vibrate when air is driven back and forth between the mouth and lungs. The result of this vibration is the production of sound.

BRONCHI

The bronchi are paired tubes that branch from the larynx and pass to each lung. Their structure is main- tained by incomplete rings of cartilage. Owing to their location cranial to the coelom, they are very difficult to observe by dissection. Therefore, observe them in Figure 5.2 and diagrammatically in Figure 6.1 rather than through dissection.

LUNGS

To observe the lungs in the cranial end of the coelom, extend your ventral midsection cut through the pectoral girdle if you have not already done so. Now cut tranversely toward the corner of the jaw on each side and pull back the cut ends of the body wall. The paired lungs are located dorsal to the liver and heart, so carefully push them to one side. If the lungs in your frog are deflated and difficult to see, insert a blowpipe through the larynx and blow gently. This will inflate the lungs as it would during inspiration. Now remove one lung and section it. Notice that it is internally divided into numerous separate partitions.

The Circulatory System

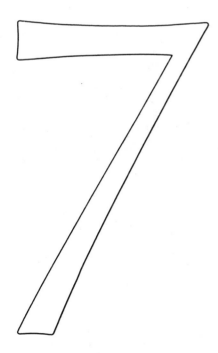

THE PRIMARY FUNCTION of the circulatory system is to transport substances including oxygen, carbon dioxide, nutrients, metabolic waste products, and hormones throughout the body. Structurally, it consists of a vast network of blood vessels that carry these substances away from (via **arteries**) and back toward (via **veins**) the central pumping organ, the **heart**.

HEART

Your study of the frog heart will begin with an examination of its external features and associated blood vessels. This will be followed by removal of the heart and its dissection to reveal internal features.

EXTERNAL FEATURES OF THE HEART

Locate the heart once again in your specimen (Fig. 4.1). Notice that it is enclosed within a membranous sac. This sac is called the **pericardium** and separates the heart from the coelom. With a pair of fine pointed scissors, cut the pericardium down the center. The space between the pericardium and the heart is the **pericardial cavity**. Pull back the pericardium to observe the ventral surface of the heart (Fig. 7.1).

With the pericardium peeled away, examine the general shape of the heart. The pointed, caudal end is called the **apex**, and the somewhat flattened, cranial end is the **base**. The two major divisions of the heart can now be identified externally:

Ventricle: the caudal half of the heart which forms the apex. In the frog it is a single, muscular chamber, but in higher vertebrates it is divided in half to form two chambers, right and left.
Atria: the two chambers that form the cranial half of the heart. The atria are generally darker in color and are composed of thinner walls than the ventricle.

BLOOD VESSELS OF THE HEART

You are now ready to study the major blood vessels of the heart. Most preserved specimens have had colored

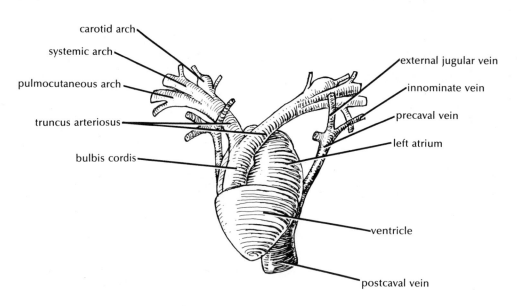

FIGURE 7.1. The heart, ventral view

material (usually latex) injected into some of their vessels after death. This color coding distinguishes between arteries and veins and makes the blood vessels tougher and more elastic. Specimens that have been doubly injected have a red material in their arteries and a blue material in their veins. Keep this coding in mind as you identify the vessels below.

Notice that the walls of the frog heart do not contain arteries or veins. In higher vertebrates, the heart wall receives oxygen and nutrients by a continuous flow of blood carried within blood vessels located within the heart wall called **coronary arteries**, and waste is removed by **coronary veins**. In the frog, the heart wall is nourished by the blood inside the heart chambers. This is possible because the walls of the frog heart are much thinner than that of higher vertebrates in order to permit diffusion, and all chambers are in direct contact with oxygenated blood.

Using Figure 7.1 as reference, examine the major vessels of the heart on the ventral side first.

Bulbis cordis: a large artery that arises from the right side of the ventricle. Because it contracts during each normal heart cycle, it is often considered as an accessory chamber of the heart.

Truncus arteriosus: a cranial continuation of the bulbis cordis which contains two lateral branches, right and left. Each lateral branch further divides into three smaller vessels: the **carotid arch**, the **systemic arch**, and the **pulmocutaneous arch**.

Sinus venosus: Carefully lift the heart away from its resting place and, with a scalpel, cut through all vessels that hold it to the body approximately one-half inch from their union with the heart. Now turn the heart over to expose the dorsal surface (Fig. 7.2). The thin-walled sac in the center of the dorsal surface is the sinus venosus. It is formed by the union of three veins: the single **postcaval vein** or **caudal vena cava** and the right and left **precaval veins** or **cranial vena cavae**. The sinus venosus transports deoxygenated blood into the right atrium.

Pulmonary vein: located on the dorsal side just cranial to the left cranial vena cava. It carries blood newly oxygenated from the lungs to the left atrium.

INTERNAL FEATURES OF THE HEART

If your frog heart was obtained from an injected specimen, it will be internally clogged with latex. The presence of latex will make it impractical to dissect the heart, so it is advisable to obtain a heart that originated from a noninjected preserved specimen or a recently living specimen. At this point, your instructor may elect to dissect a heart from a living frog as a demonstration.

To dissect the heart, insert your scissors into the wall of the ventricle at the apex. Carefully cut through the heart wall in a cranial direction so that the heart will be halved along the frontal plane. Make small, shallow cuts as you proceed until the walls of the ventricle and atria are removed. Now insert one point of your scissors into the opening between the ventricle and the bulbis cordis. Make a ventral cut in the bulbis cordis, trimming the edges as necessary to expose the internal structure. Refer often to Figure 7.3, and identify the following internal structures:

Ventricle: Note the thickness of the ventricular wall. The bulk of this wall is composed of cardiac muscle. Also note the muscular ridges called **trabeculae**. The ventri-

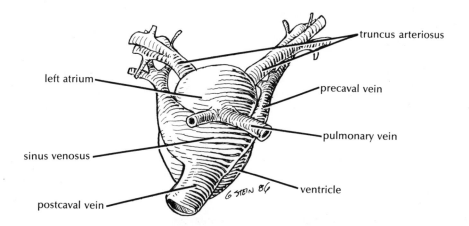

left atrium

truncus arteriosus

precaval vein

pulmonary vein

sinus venosus

ventricle

postcaval vein

G STEIN 86

FIGURE 7.2.   The heart, dorsal view

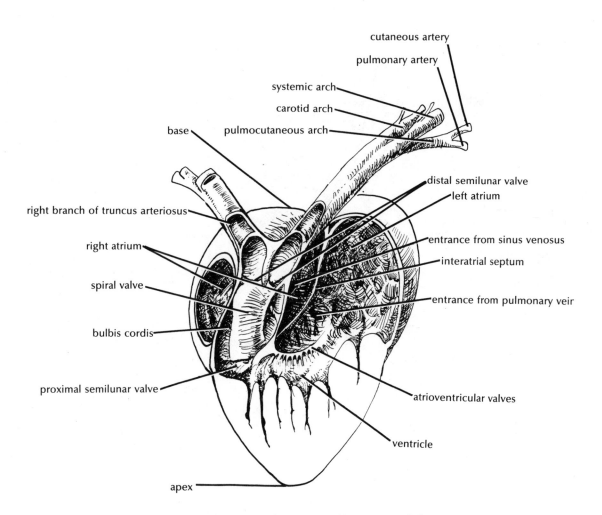

cutaneous artery

pulmonary artery

systemic arch

carotid arch

pulmocutaneous arch

base

distal semilunar valve

left atrium

right branch of truncus arteriosus

entrance from sinus venosus

right atrium

interatrial septum

spiral valve

entrance from pulmonary veir

bulbis cordis

proximal semilunar valve

atrioventricular valves

ventricle

apex

FIGURE 7.3.   Frontal section of heart, ventral view

cle receives blood from the atria through four one-way valves called **atrioventricular valves**. Its contraction propels blood through three one-way valves called **semilunar valves** into the bulbis cordis.

Bulbis cordis: internally, it extends from the proximal semilunar valves to the distal semilunar valves that open into the truncus arteriosus. In the center of the bulbis cordis attached to its dorsal wall is a **spiral valve**. This structure controls the direction of blood flow into the carotid, systemic, and pulmocutaneous arches.

Atria: The right and left atria are completely separated by the **interatrial septum**. Locate the opening of the sinus venosus into the right atrium and the opening of the pulmonary vein into the left atrium. The sinus venosus carries a mixture of deoxygenated blood from the body plus oxygenated blood from the skin and mouth. At its entrance into the right atrium are two valves that prevent blood from flowing back into the sinus venosus when the atrium contracts. The pulmonary vein carries oxygenated blood from both lungs into the left atrium. Con-

traction of the atria will, therefore, propel both oxygenated and deoxygenated blood (mixed blood) into the ventricle.

BLOOD VESSELS CRANIAL TO THE HEART

In your study of the vascular system of the frog, you will examine cranial blood vessels first. As you dissect the blood vessels, bear in mind that they are subject to considerable variation and may be a bit different in each specimen. The diagrams represent the most common routes encountered.

In order to locate the vessels, observe the frog's ventral side and push aside the organs in the cranial pleuroperitoneal cavity. For a more complete view, you may elect to remove the right forelimb. To do this, cut through the right shoulder muscles and through the shoulder joint completely. Take care to avoid cutting through vessels unnecessarily. Also remove the posterior portion of the right jaw with bone clippers. This will provide a more unobstructed view of the cranial vessels.

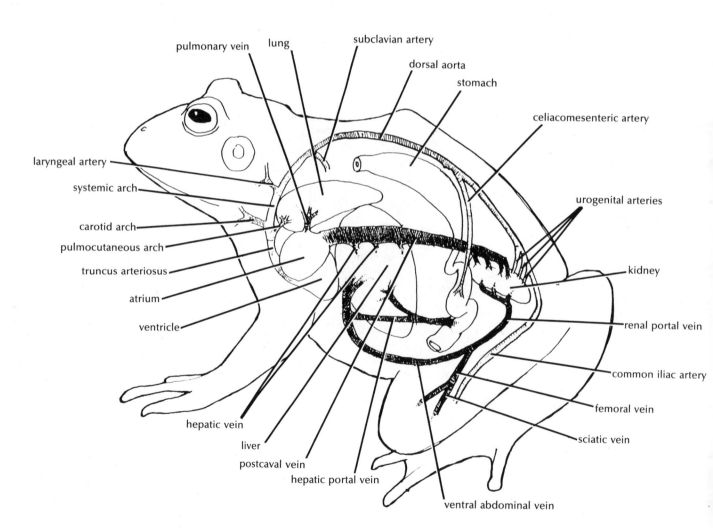

FIGURE 7.4.   Schematic diagram of circulatory pathways

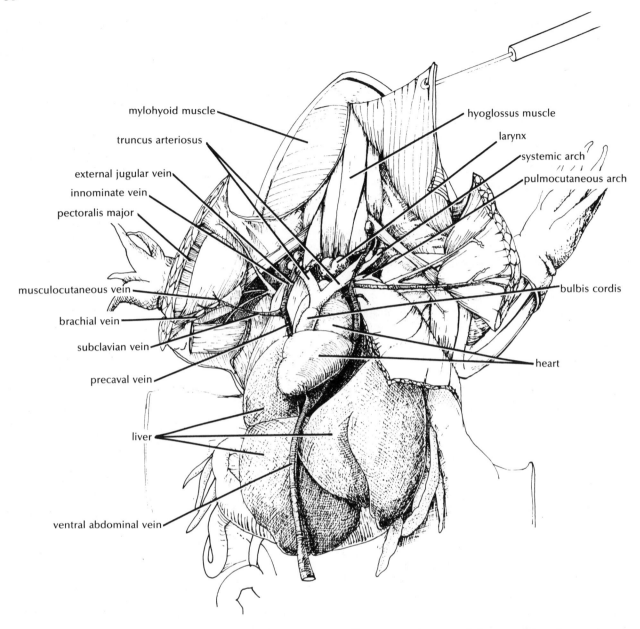

mylohyoid muscle

truncus arteriosus

external jugular vein

innominate vein

pectoralis major

musculocutaneous vein

brachial vein

subclavian vein

precaval vein

liver

ventral abdominal vein

hyoglossus muscle

larynx

systemic arch

pulmocutaneous arch

bulbis cordis

heart

FIGURE 7.5.   Cranial blood vessels, ventral view. All organs are intact, and the musculature is separated to show peripheral vessels.

ARTERIES

Cranial arteries are tributaries of the truncus arteriosus. As indicated, the truncus arteriosus divides, or **bifurcates**, into right and left branches, each of which further bifurcate into the carotid arch, the systemic arch, and the pulmocutaneous arch. Locate the origin of each arch and trace their branches as follows (Figs. 7.4, 7.5, 7.6):

Carotid arch (paired): the cranial branch from the truncus arteriosus. It passes cranially toward the head on each side. The carotid arch contains the following major branches:

External carotid artery (paired): the first branch from the carotid arch. It supplies the tongue and lower jaw.

Internal carotid artery (paired): the continuation of the carotid arch past its junction with the external carotid. It supplies the brain, eye, and upper jaw by way of the **cerebral**, **opthalmic**, and **palatine arteries**.

Systemic arch: the middle branch of the truncus arteriosus. It passes dorsally to one side of the esophagus and continues caudally until it unites with the systemic arch from the opposite side at a point cranial to the kidneys.

The unification of the two arches at the dorsal midline marks the origin of the **dorsal aorta**. Each systemic arch gives rise to the following branches:

Laryngeal artery (paired): the first branch from the systemic arch. It is a small vessel that passes dorsally to supply the pharynx and larynx.

Occipitovertebral artery (paired): extends dorsally from the systemic arch to supply the muscles of the jaw and back.

Subclavian artery (paired): arising from the systemic arch just caudal to the origin of the occipitovertebral artery. The subclavian passes to the shoulder and forearm to supply these regions.

Pulmocutaneous arch (paired): the caudal branch from the truncus arteriosus. It passes dorsally a short distance before it bifurcates into the **pulmonary** and **cutaneous arteries** on each side.

Pulmonary artery (paired): From its origin at the pulmocutaneous arch, it passes caudally to supply the lung on each side.

Cutaneous artery (paired): It passes dorsally from the pulmocutaneous arch before it branches extensively to supply the skin.

VEINS

The final destination of blood flowing through veins is the atria of the heart, as veins carry blood toward the heart. Because of the difficulty in tracing veins from the periphery to the heart in concordance with blood flow, the protocol below asks you to trace the veins beginning at the heart as you have done with the arteries. Using Figures 7.4 through 7.6, locate the following major cranial veins in your specimen:

Precaval vein (paired): also called the **cranial vena cava**. It receives blood from the cranial region of the body by way of three veins: the **external jugular**, the **innominate**, and the **subclavian**. The right and left precaval veins convey blood into the sinus venosus.

External jugular vein (paired): It receives blood from the **mandibular vein**, which drains the lower jaw and the **lingual vein**, which drains the tongue and floor of the mouth. The external jugular vein may be viewed as it extends caudally to the precaval vein.

Innominate vein (paired): Locate the innominate vein as the middle junction with the precaval vein. It is formed by the union of the **internal jugular vein**, which drains blood from the brain, and the **subscapular vein**, which drains the shoulder and dorsal part of the forelimb.

Subclavian vein (paired): a large vein that drains the forelimb and forms the caudal junction with the precaval vein. It is formed by the union of the **brachial vein** and the **musculocutaneous vein**.

BLOOD VESSELS CAUDAL TO THE HEART

In order to view the caudal blood vessels, push aside the visceral organs as you proceed through the protocol below. If you have time, you may remove structures to obtain a more unobstructed view. If this is the case, begin by removing the liver.

Starting at the cranial end of the liver on the right side, free it from the heart and lungs and remove the right lobe, but take care not to damage the **postcaval vein** (below) while you do this. If your specimen is a female, be careful to preserve all portions of the reproductive system. Now remove the left lobes of the liver while carefully avoiding damage to the postcaval vein.

To remove the gastrointestinal tract, begin by cutting the large intestine near the urinary bladder. Now cut the mesentery to free the intestine and stomach, and cut the stomach near its union with the esophagus. Remove the stomach and intestines, and trim any remaining portions of the mesentery and peritoneum to make a clean dissection of the kidneys, gonads, and blood vessels against the dorsal body wall. Identify the following caudal blood vessels, and refer often to Figures 7.4 through 7.8:

ARTERIES

Caudal arteries are tributaries of the left and right systemic arches, which converge after passing caudally to form the **dorsal aorta**. Locate this junction on the dorsal body wall of your specimen, and trace the following branches of the dorsal aorta (Fig. 7.8):

Celiacomesenteric artery: a large vessel that branches from the dorsal aorta near its origin. It passes ventrally to supply the visceral organs by way of two major tributaries: the **celiac artery**, which supplies the liver and stomach, and the **cranial mesenteric artery**, which supplies the small and large intestines.

Urogenital arteries: six pairs of vessels that branch from the dorsal aorta to supply the kidneys, gonads, and fat bodies.

Lumbar arteries: several arteries that pass from the dorsal aorta into the muscles of the dorsal body wall to supply them and the spinal cord with blood.

Caudal mesenteric artery: a single vessel that emerges from the caudal portion of the dorsal aorta and extends to the large intestine.

Common iliac artery (paired): The dorsal aorta bifurcates into the right and left common iliacs in the pelvic region. Each common iliac contains the following branches as it extends caudally:

Hypogastric artery (paired): supplies the urinary bladder.

Epigastric artery (paired): supplies the ventral body wall.

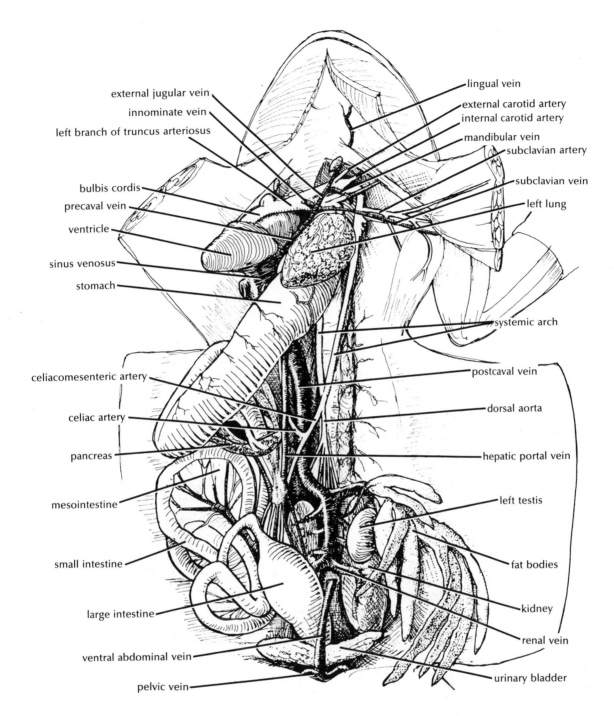

external jugular vein
innominate vein
left branch of truncus arteriosus

bulbis cordis
precaval vein
ventricle
sinus venosus
stomach

celiacomesenteric artery

celiac artery

pancreas

mesointestine

small intestine

large intestine

ventral abdominal vein

pelvic vein

lingual vein
external carotid artery
internal carotid artery
mandibular vein
subclavian artery
subclavian vein
left lung

systemic arch

postcaval vein

dorsal aorta

hepatic portal vein

left testis

fat bodies

kidney

renal vein

urinary bladder

FIGURE 7.6. Blood vessels of the coelom, ventral view. The liver is removed, and the remaining visceral organs are reflected to the right side in this example of a male specimen.

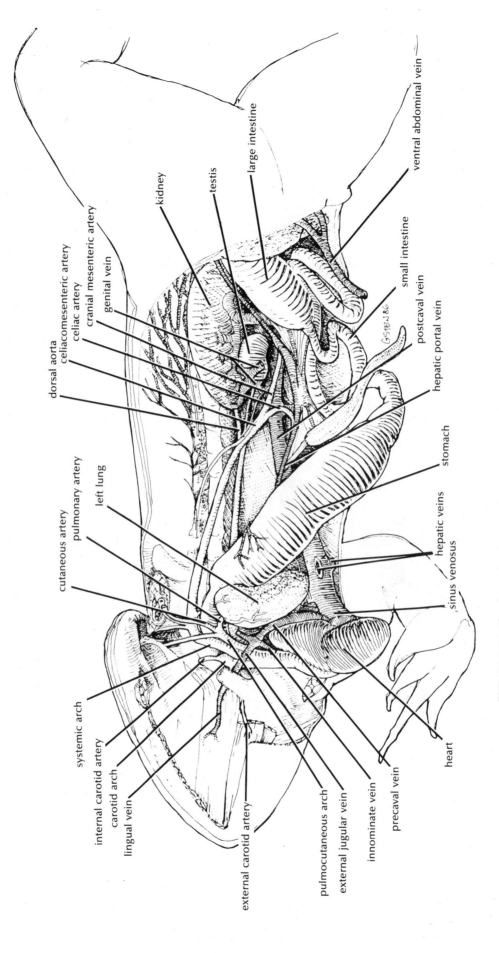

cutaneous artery
pulmonary artery
left lung

dorsal aorta
celiacomesenteric artery
celiac artery
cranial mesenteric artery
genital vein

kidney
testis
large intestine

ventral abdominal vein

small intestine
postcaval vein
hepatic portal vein

GSTEIN 86

stomach

hepatic veins

sinus venosus

systemic arch

internal carotid artery
carotid arch
lingual vein

external carotid artery

pulmocutaneous arch
external jugular vein
innominate vein
precaval vein

heart

FIGURE 7.7. Blood vessels of the coelom, ventrolateral view of the left side with the liver removed. The left forelimb is also removed in this male.

Femoral artery (paired): supplies the hip and thigh regions.

Sciatic artery (paired): the caudal continuation of the common iliac. It supplies the thigh, and its caudal branches supply the shank and foot.

VEINS

The sinus venosus collects all venous blood coming from the body caudal to the heart. Locate this large vein once more, and identify the following veins in sequence from the sinus venosus to the body periphery:

Postcaval vein: also called the **caudal vena cava**. It is a single, large vein and is the only vessel that brings caudal venous blood to the sinus venosus. It arises between the kidneys where it is formed by the union of the four pairs of **renal veins** and receives the following vessels:

Renal veins: four pairs of veins, two from each kidney, which merge to form the postcaval vein. They drain the kidneys of venous blood.

Genital veins (paired): arise from the testes or ovaries.

Hepatic veins (paired): arise from the liver to join the postcaval vein near its union with the sinus venosus.

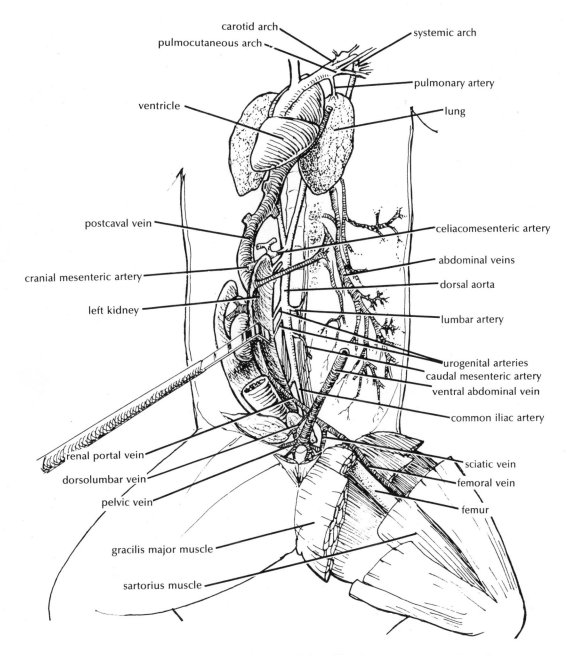

FIGURE 7.8. Coelomic blood vessels, ventral view. The digestive organs in this male are removed, and the urogenital organs of the right side are reflected medially.

Ventral abdominal vein: a single, large vein that you encountered when the abdominal wall was first opened. It extends from the pelvic region to the liver, where it divides into right and left branches that enter the right and left lobes of the liver.

Hepatic portal vein: a single vein that extends between the G.I. tract and the liver. It is termed *portal* because it does not pass directly to the heart but instead enters an organ and subdivides within it to join its capillary system. In the frog, there are two portal systems, the hepatic and the renal (below). In higher vertebrates, the hepatic portal system is the only one present.

Renal portal vein (paired): located along the lateral edge of each kidney. As it nears the kidney, it branches into numerous tributaries that extend into the kidney. It is formed caudally by the union of the femoral and sciatic veins.

Dorsolumbar vein (paired): a small vein that drains the dorsal body wall and passes into the renal portal vein.

Femoral vein (paired): the major vein that drains the hind limb. It passes through muscles of the thigh to the pelvic region where it merges to help form the caudal end of the renal portal vein.

Pelvic vein (paired): The right and left pelvic veins extend from their union with the femoral veins to unite along the midventral line. Their mergence forms the ventral abdominal vein.

Sciatic vein (paired): drains the dorsal portion of the thigh. The sciatic vein merges with the femoral vein in the pelvic region on each side to form the right and left renal portal veins.

The Urogenital System

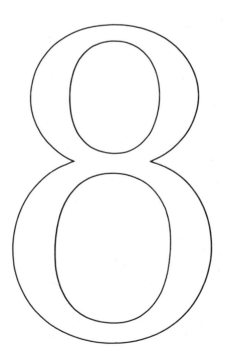

THE UROGENITAL SYSTEM has two primary functions that are quite different from each other: the formation and elimination of urine, and reproduction. These two functions are combined in one common system in amphibians because of the common location and close association of a number of their organs. As you will discover, several urinary organs lie in direct contact with reproductive structures, and in some cases both functions take place within one organ. In mammals, urinary and reproductive functions are more anatomically separate, so these functions are often separated into two distinct systems.

The urinary functions within the urogenital system primarily involve the **kidneys**. These functions include the removal of nitrogen-containing waste products from the bloodstream and their transport out of the body in the form of liquid urine, maintenance of the osmotic balance of fluids, and the control of blood cell formation in red bone marrow.

In sharp contrast, the reproductive functions of the urogenital system mainly involve the **gonads**, organs that produce the sex cells, or **gametes**. In the female, the reproductive organs are highly adapted for the production, storage, and release of the female gametes (**ova**). In the male, the reproductive organs are adapted for the production of the male gametes (**spermatozoa**) and their release onto the newly released ova, or eggs, for their fertilization.

In this chapter, dissection of the urogenital system is divided into the urinary portion and the reproductive portion. The reproductive portion is further divided into a discussion of the male and female systems.

URINARY ORGANS

To locate the urinary organs against the dorsal body wall, the visceral organs must be removed. Follow the procedure described on page 34 to remove the liver, stomach, and intestine if you have not already done so. Now identify the following organs of the urinary tract:

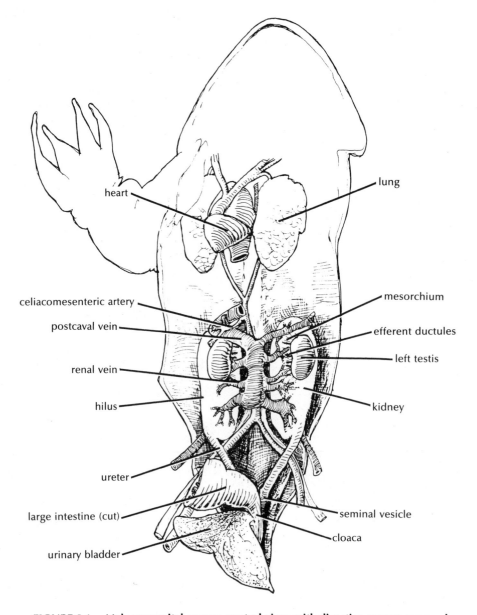

FIGURE 8.1.   Male urogenital organs, ventral view, with digestive organs removed

KIDNEYS

The kidneys (paired) are elongated, reddish-brown organs that lie against the dorsal body wall (Fig. 8.1). They lie dorsal to the peritoneum and are covered by it only on the ventral surface. Internally they are composed of many microscopic tubule arrangements called **nephrons**,[1] which are in close association with extensive capillary networks. The nephrons are the site of functional activities that occur within the kidneys. The urine that has been formed within the nephrons will eventually pass to collecting ducts. In male frogs these collecting ducts also receive tubes that carry sperm from the testes. In both sexes, the collecting ducts empty into the **ureters**.

Examine the surface of the kidney closely without disturbing its position. The medial border faces the opposite kidney and is straight, while the lateral border is convex. The convex border is called the **hilus** and is where the ureter originates. Note the yellowish, diffuse structure on

[1]The correct plural form of the word *nephron* is actually *nephroi*. *Nephrons* is used here because it is the most common usage.

the ventral side of the kidney. This is the **adrenal gland**, an endocrine gland that secretes the hormones, epinephrine and cortisol, into the bloodstream.

URETERS

The paired ureters, or **Wolffian ducts**, are conducting tubes that transport urine and in the male, sperm, from the hili of the kidneys caudally to the cloaca. They are small tubes and are located against the dorsal body wall.

URINARY BLADDER

The urinary bladder is a thin-walled sac that is attached to the ventral side of the cloaca with which it unites. It is a temporary storage site for urine that it receives from the ureters.

CLOACA

The cloaca is a short tube that extends from its union with the caudal end of the large intestine to the exterior. It is a common chamber that receives feces, urine, and gametes and releases them to the exterior through the **cloacal aperture**. To observe it, make a cut with a scalpel through the pelvic girdle on the ventral side along the midline and pull the hind legs back. Cut the urinary bladder near its union with the cloaca and remove it. With the cloaca now exposed ventrally, make an incision through its wall and pull back the flaps to view its internal structure. Locate the openings of the ureters and the urinary bladder into the cloaca.

REPRODUCTIVE ORGANS

The reproductive organs vary in location and structure between males and females, so the descriptions are separated below. Follow the protocol for the sex of your specimen first. If available, examine a specimen of the opposite sex from your lab partner or a demonstration specimen as well.

MALE REPRODUCTIVE ORGANS

Locate the following organs of the male reproductive tract (Fig. 8.1):

Testes. The testes are a pair of oval structures lying at the cranial portion of the kidneys on the ventral side. They are the male gonads; as such they produce the male gametes, spermatozoa. Each testis is anchored to the kidney dorsal to it by a peritoneal membrane called the **mesorchium**, which extends from the medial border of the testis.

Efferent Ductules. The efferent ductules are numerous small tubes that extend from each testis into the kidney dorsal to it. Inside the substance of the kidney they merge

with the **transverse collecting tubules**, which receive urine as it is formed from the nephrons in addition to sperm from the efferent ductules. Still within the kidney, the collecting tubules converge to form the ureter, which exits the kidney and passes to the cloaca.

Ureters. The male ureter is sometimes called the **ductus deferens**, because it is a tube that transports sperm away from the testis as well as urine from the kidney. Near its union with the cloaca, the ureter contains a thickening of its wall. This is caused by the presence of a gland, the **seminal vesicle**, which secretes fluid aiding in the transport of sperm.

Cloaca. As a male reproductive organ, the cloaca serves as a temporary reservoir for sperm prior to fertilization. Fertilization is performed externally in frogs once a year during the mating season, which is normally in the spring. During this process, the male mounts the back of a female. A firm grip is established by the male with the use of the thumbs (first digits), which are considerably larger than those of the female and are therefore adapted for this purpose. The male remains in this position until the female's eggs are released. As the eggs exit her cloaca, the male's sperm is released. Fertilization is therefore external.

Fat Bodies. The fat bodies are not organs but yellow clusters of fat tissue located cranial to the kidneys against the dorsal body wall. They provide energy storage that can be called upon during the mating season or dormancy. They are present in both males and females.

FEMALE REPRODUCTIVE ORGANS

Locate the following organs of the female reproductive tract (Fig. 8.2):

Ovaries. The paired ovaries are located ventral to each kidney. They exhibit a marked variation in size according to the season of the year. If your specimen was collected during the late fall, the ovaries appear as small membrane-bound organs. If the female you are studying was collected during springtime, the ovaries appear enlarged and occupy a large portion of the coelom. Note the division of each ovary into several compartments. Each is covered by a fold of the peritoneum that anchors it to the dorsal body wall and is called the **mesovarium**. During ovulation, the eggs within these compartments break through the mesovarium to enter the coelom.

Oviducts. The paired oviducts are small, coiled tubes that extend from the cranial end of the coelom to the **uterus** near the caudal end of the coelom. At their origin at the cranial end of the coelom, where they may be observed lateral to the heart, is an expansion called the **ostium**, (plural form is **ostia**) which helps channel eggs into

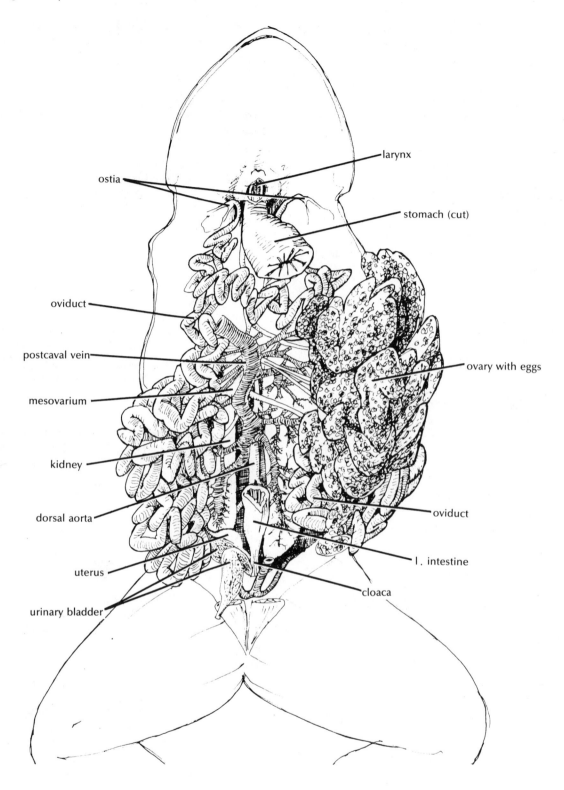

FIGURE 8.2.   Female urogenital organs, ventral view, with digestive organs removed. The ovaries and oviducts are in an enlarged state, as the specimen was collected during spring mating season.

the oviduct after their expulsion into the coelom during ovulation. The oviducts transport eggs from the coelom to the uterus.

Uterus. In frogs, the uterus, or **ovisac**, is a paired structure. Each uterus may be viewed at the caudal end of the coelom, where it appears as a pear-shaped expansion of the oviduct. The uteri open into the cloaca ventral to the opening of the ureters.

Cloaca. The reproductive function of the female cloaca is as a temporary storage site for eggs as they pass between the uterus and the exterior. By the time they are ready to be released to the exterior, they have become surrounded by a gelatinous material secreted while in the oviduct. When the jelly comes in contact with water, it swells to form a transparent coat that protects the egg from hazards of the external environment.

The Nervous System

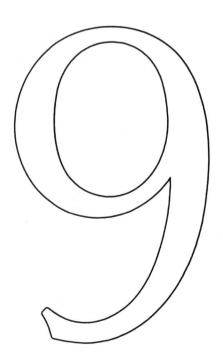

T HE PRIMARY FUNCTION of the nervous system as a whole is to maintain the body in a relatively stable condition, or **homeostasis**, despite fluctuating environmental conditions. This is accomplished by controlling body activities through the generation of nerve impulses that travel along the nerve cells or **neurons**. This means of body control is rapid and specific to the region that receives the impulse.

The nervous system is not the only body system that functions in maintaining internal stability; the **endocrine system** also plays an important role in homeostasis. In contrast to the rapid and specific effects of nervous control, endocrine control is by way of the secretion of **hormones** that are carried more sluggishly by the circulatory system. Hormonal stimulation thus results in slower effects that may affect different parts of the body. Hormones are secreted by glands that are widely scattered, making dissection of the endocrine system as a whole very difficult. The endocrine system will not be studied in one place, but endocrine glands are presented as they are observed in the course of dissection.

In this chapter, the nervous system is presented, and its complete dissection is described. Its divisions into the **central nervous system** and the **peripheral nervous system** are developed below.

THE CENTRAL NERVOUS SYSTEM

The central nervous system (CNS) consists of the brain and the spinal cord. It integrates, regulates, and controls the body's activities and relays impulses between the brain and the peripheral nerves. It is a hollow structure; through it runs a **central canal** that expands in certain regions of the brain to form **ventricles**. Within these cavities and also surrounding the spinal cord and brain is **cerebrospinal fluid**, a slowly circulating colorless fluid that provides nourishment and a protective liquid cushion for the central nervous system.

BRAIN

The brain is the center for integration, regulation, and control of all body activities. It is located within the well-protected **cranial cavity** of the skull. Owing to this location it is the most challenging organ to dissect. The protocol outlined below describes the dissection of the brain and spinal cord together to enable you to conserve time. Follow it closely when you are ready to begin.

1. Remove the skin, muscles, and connective tissue from the dorsal surface of the skull.
2. Remove the skin, muscles, and connective tissue from the dorsal surface of the vertebral column. Consult Figure 2.1 to prevent damaging the spinal cord that lies within the column.
3. Place your specimen on its back in a dissecting tray containing enough 3 percent nitric acid to cover completely the exposed skull and vertebral column.[1] Leave the frog in this position for two to three days; this will dissolve the mineral salts that provide hardness to the bone and will leave mostly soft cartilage.
4. Remove your frog from the acid bath and rinse it in running water. Return it to a clean dissecting tray and place it on its ventral side. Bend the head down slightly and insert one point of your scissors into the foramen magnum at the base of the skull—don't penetrate too deeply! Cut forward along the dorsal surface of the skull as far rostral as is reasonable.
5. With forceps and scissors, carefully chip away the skull from the midline to the otic capsule surrounding the tympanum. Note the tough membrane adhering to the inner surface of the cranium. This is the **dura mater**, which is one of the two layers of **meninges** surrounding the brain and spinal cord. With the brain exposed, you can observe the thinner **pia mater** that lies close to the surface of both the brain and spinal cord.
6. To remove the brain, gently raise the anterior part with small forceps first. As it is lifted, cut the cranial nerves close to the skull. Continue working in this way until the brain is completely free. Cut the spinal cord and remove the brain from the cranial cavity.
7. With extreme care, insert the point of your scissors into the vertebral canal and cut in a caudal direction to remove the dorsal surface of the vertebral column. Make this cut shallow, and work slowly to avoid damaging the spinal cord and nerves. Now cut away the lateral portions of the vertebral column that remain. Owing to its many connections, the cord will not be removed as was the brain. Instead, observe it from its exposed dorsal surface.

---

[1]To avoid using acid to soften the skull, small bone forceps may be used to crack through the bone. This is not recommended if the brain is to be removed without damage.

The brain of the frog is similar to that of other vertebrates in that it consists of three primary sections, the **forebrain**, the **midbrain**, and the **hindbrain**. Identify the following features of the brain on your specimen (Figs. 9.1 through 9.3). Also note the approximate positions of the ten cranial nerves as listed in Table 2.

Forebrain: Also called the **prosencephalon**, it is the largest section of the brain. It occupies the anterior and middle regions and consists of the **cerebral hemispheres** and the **diencephalon**.

　Cerebral hemispheres: Also called the **telencephalon** or **cerebrum**, they consist of right and left portions separated by a median **longitudinal fissure**. At the anterior end of each hemisphere is a swelling called the **olfactory lobe**, which receives olfactory information from the nasal epithelium.

　Diencephalon: located posterior to the cerebral hemispheres. It is internally divided into a **thalamus** and a smaller **hypothalamus** (not shown). In the center of the diencephalon is a small stalk that, before removal, was attached to a small **pineal body** (**epiphysis**). The pineal body is a rudimentary third eye that was functional in early vertebrates and lost in the course of evolution. On the dorsal surface, two large swellings on the diencephalon are the **optic lobes**, which receive visual information from the eyes. On the ventral surface, observe the crossing of the optic nerves from the eyes, called the **optic chiasma**. Also on the ventral surface may be the **pituitary gland** (or **hypophysis**) if it was not removed accidentally during removal of the brain. This is an important endocrine gland and is anchored to the diencephalon by a stalk posterior to the optic chiasma, called the **infundibulum**.

Midbrain: also called the **mesencephalon**. It occupies a region posterior to the diencephalon. It can only be observed by sectioning the brain and, thus, is not shown.

Hindbrain: also called the **rhombencephalon**. It is the posterior portion of the brain that unites with the spinal cord. Locate the **cerebellum** (the visible part of the **metencephalon**), which is posterior to the optic lobes on the ventral surface. Also identify the **medulla oblongata** (myelencephalon), which tapers to merge with the spinal cord.

SPINAL CORD

The spinal cord transmits nerve impulses to and from the brain. Structurally, it is a semicylindrical mass of nerve tissue that is surrounded by the meninges. Beginning cranially as a caudal continuation of the medulla oblongata, it passes through the vertebral canal and terminates as a slender filament, the **filum terminale**, in the pelvic region. In the brachial and lumbar regions it con-

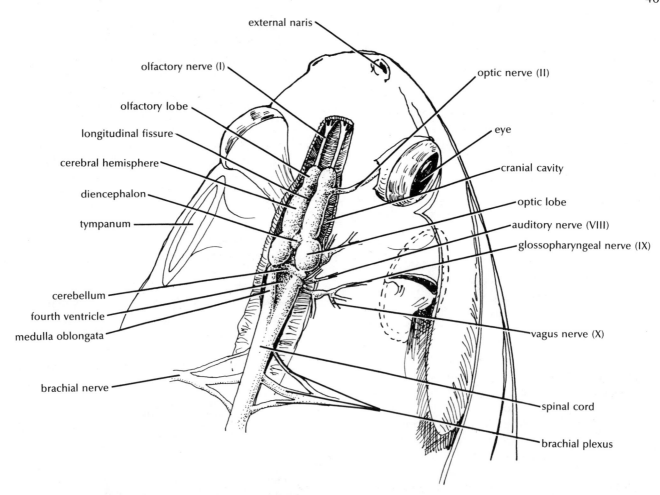

FIGURE 9.1. The brain within the cranial cavity, dorsal view. Extrinsic muscles of the right eye and the right tympanum are removed.

TABLE 2. The Cranial Nerves

Cranial Nerve	Point of Union with Brain	Innervation	Sensation or Action
Olfactory (I)	Olfactory lobe	Epithelium of nasal cavity	Sense of smell
Optic (II)	Thalamus	Sensory cells in retina of eye	Sense of vision
Oculomotor (III)	Ventral surface of midbrain	Muscles of the eye	Movement of eye, control of light entering eye
Trochlear (IV)	Medulla, ventral surface of cerebellum	Muscle of eyeball	Rotation of eyeball
Trigeminal (V)	Lateral portion of medulla	Eyeball, skin of skull, jaw, teeth, mouth, and tongue	Sensations on skull, eyeball, and mouth; movement of jaw
Abducens (VI)	Ventral surface of medulla	Muscle of eyeball	Movement of eyeball
Facial (VII)	Lateral segment of medulla	Tongue, face, jaw muscles	Sense of taste; muscle movement for mastication
Auditory (VIII)	Lateral segment of medulla	Sensory cells in inner ear	Sense of hearing and equilibrium
Glossopharyngeal (IX)	Lateral segment of medulla	Pharynx, caudal third of tongue	Sense of taste; sensation of throat region
Vagus (X)	Lateral segment of medulla	Muscles of the pharynx, larynx, lungs, heart, diaphragm, and viscera; sensory cells of larynx, viscera	Contraction of innervated muscles; sensation of innervated regions

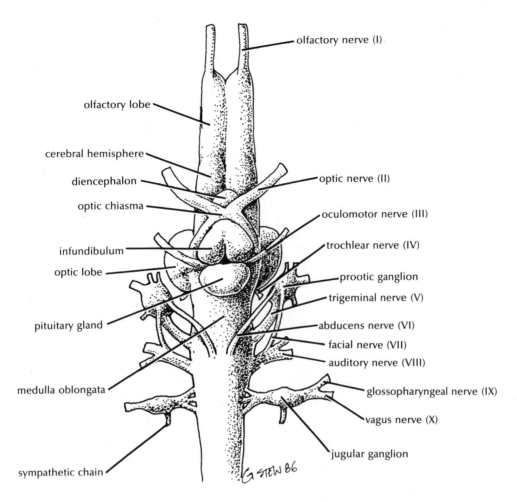

olfactory nerve (I)

olfactory lobe

cerebral hemisphere

diencephalon

optic chiasma

optic nerve (II)

oculomotor nerve (III)

trochlear nerve (IV)

infundibulum

optic lobe

prootic ganglion

trigeminal nerve (V)

pituitary gland

abducens nerve (VI)

facial nerve (VII)

auditory nerve (VIII)

medulla oblongata

glossopharyngeal nerve (IX)

vagus nerve (X)

jugular ganglion

sympathetic chain

FIGURE 9.2.   The brain and cranial nerves, ventral view

tains swellings, called **enlargements**, that give rise to the nerves that supply the appendages.

A pair of **spinal nerves** arises from each of ten segments and exits the vertebral canal. They are named according to the region of the vertebral column where they arise.

Examine the spinal cord of your specimen and identify the components described above and indicated in Figure 9.3.

THE PERIPHERAL NERVOUS SYSTEM

The peripheral nervous system (PNS) consists of nerves located outside the CNS which transmit nerve impulses to and from the spinal cord and brain. Nerves that conduct impulses toward the CNS are termed **afferent**, or **sensory**, and those that conduct impulses away from the CNS are **efferent**, or **motor**.

The PNS is functionally divided into a portion that is involved with voluntary activities and one involved with involuntary activities. All nerves that carry involuntary

impulses belong to the subdivision of the PNS called the **autonomic nervous system**. The autonomic nervous system is further divided into a **sympathetic division**, which is activated mainly during emergency or stressful situations, and a **parasympathetic division**, which is active during moments of rest and repose. PNS nerves that carry voluntary information are referred to as afferent or efferent, depending on the direction of impulse flow.

Voluntary and involuntary portions of the PNS are structurally separated. The voluntary portion consists of segmented spinal nerves that arise from the spinal cord. Some of these nerves unite with each other in the musculature soon after exiting the vertebral column to form networks, or **plexi**. From the plexi emerge nerves that supply mainly the appendages. In the frog this is true of all spinal nerves except Nos. 4, 5, and 6, which exit the spinal cord to pass directly to the abdominal skin and muscles.

The involuntary portion of the PNS, the autonomic system, contains nerves that also exit the spinal cord

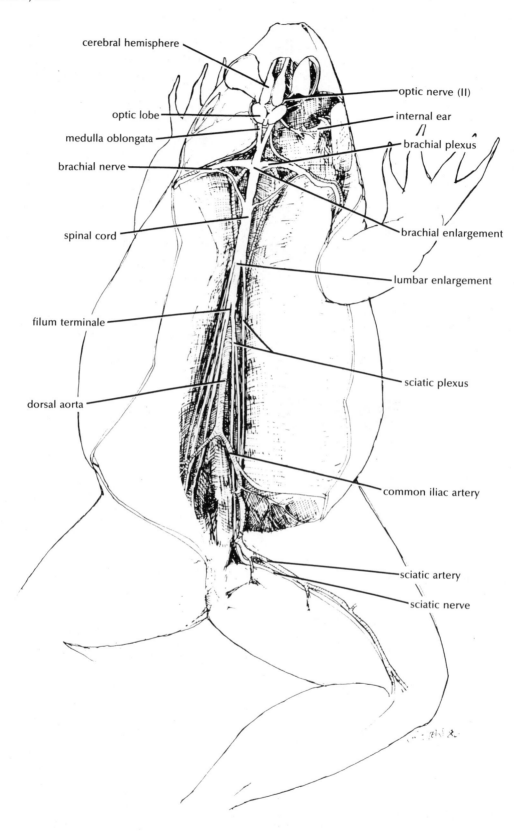

cerebral hemisphere

optic nerve (II)

optic lobe

internal ear

medulla oblongata

brachial plexus

brachial nerve

brachial enlargement

spinal cord

lumbar enlargement

filum terminale

sciatic plexus

dorsal aorta

common iliac artery

sciatic artery

sciatic nerve

**FIGURE 9.3.   The nervous system, dorsal view. The cranium and spinal column
have been removed.**

within spinal nerves. Instead of forming plexi or passing to the abdominal region, however, the nerves exit the spinal nerves near the vertebral column to form nerve clusters, or **ganglia**. From the ganglia the autonomic nerves pass to glands, smooth muscle, and cardiac muscle, which they innervate.

Identify the following components of the peripheral nervous system in your specimen. Refer often to Figure 9.3.

MAJOR PLEXI

In the dorsal musculature of the frog, locate the following plexi and peripheral nerves (Fig. 9.3):

Brachial plexus: located deep to the dorsalis scapula muscle of the back. It is formed by the union of the first three spinal nerves. Emerging from the brachial plexus is the large **brachial nerve**, which passes to the shoulder, where it divides into the **ulnar nerve** and the **radial nerve**. The ulnar nerve supplies flexor muscles of the forelimb and the radial nerve supplies extensor muscles of the forelimb.

Sciatic plexus: a large network located deep to the longissimus dorsi muscle. It is formed by the union of spinal nerves, Nos. 7 through 9. The largest nerve emerging from this plexus is the **sciatic nerve**, which passes down the hindleg between the iliofibularis and semimembranosus muscles of the thigh.

AUTONOMIC NERVOUS SYSTEM

The components of the autonomic nervous system that are dissectible are the ganglia and nerves of the sympathetic division, located on either side of the vertebral column. Often these white structures are damaged during previous procedures, but you should make an attempt to locate any remaining structures:

Sympathetic trunk ganglia: small, white, oval bodies that lie parallel to the vertebral column on both sides. They are vertically interconnected and are collectively referred to as the **sympathetic chain**.
Communicating rami: the small, fragile nerves that extend between the spinal nerves and the sympathetic trunk ganglia.

also by Bruce D. Wingerd and available from Johns Hopkins:

Human Anatomy and Rabbit Dissection
Rabbit Dissection Manual
Dogfish Dissection Manual
Rat Dissection Manual

The Johns Hopkins University Press

FROG DISSECTION MANUAL

This book was composed in Optima (Oracle) type by
Brushwood Graphics, Inc., from a design by Susan P. Fillion.
It was printed by Thomson-Shore, Inc., on 60-lb. Spring
Forge Offset.

NOTES

NOTES

NOTES

NOTES

NOTES